THE PREPPER'S SURVIVAL BIBLE

Sustainable Off-Grid Life-Saving Strategies to Prepare for Any Emergency Disaster Scenario With Food Stockpiling, Canning, Water Filtration and Home Defense Pro Tips

Sam Madigan

First published by Riverthane Press, 2024
First Edition Revised March 2026
ISBN 978-1-917501-01-9

Disclaimer

This book is intended to provide general guidance for survival enthusiasts and preppers. While the author and publisher have made every effort to ensure accuracy, survival situations are inherently unpredictable, and the strategies presented may not suit every individual or circumstance. Readers are strongly advised to adapt the information to their specific needs and consult manufacturer recommendations, local experts, healthcare providers, and authorities for personalized advice. This is especially important for activities like foraging, where the risk of harm from misidentification or improper preparation is significant. Proper training and additional resources should be sought before attempting foraging or other survival skills independently.

The author and publisher disclaim any responsibility for errors, omissions, injuries, or damages resulting from the use or misuse of the information in this book. Some scenarios are illustrative and created to highlight potential real-life situations. While this book was primarily written by a human, some of the content was created with the assistance of Artificial Intelligence for images and to enhance clarity and efficiency of the text. Additionally, the author and publisher do not guarantee the accuracy or reliability of external links or resources mentioned herein, and readers should exercise caution when applying information from these sources.

Furthermore, the author and publisher make no warranties regarding the suitability of this book for any specific purpose or individual. It is the reader's responsibility to evaluate the information and strategies presented and to use them at their own risk.

By using this survival guide, the reader agrees to the terms and conditions of this disclaimer.

Stay curious. Stay prepared.
But taking action is
what truly protects your
loved ones.

SAM MADIGAN

Your Path to Readiness Begins Right Here

As you dig into the pages of The Prepper's Survival Bible, I hope the strategies, knowledge, and resources within these pages become cornerstones of your survival plan. Don't forget to access your Bonus Downloadable Resources at the end of the book!

If you enjoy this book, I'd truly appreciate it if you offer some honest feedback by leaving a review. Your feedback not only helps me but also guides others in finding the tools they need to face the future with confidence.

Remember, preparedness is a journey

...and you're already ahead of the curve.

TABLE OF CONTENTS

TABLE OF CONTENTS

INTRODUCTION

> Stay curious. Stay prepared.
> But taking action is
> what truly protects your
> loved ones.
>
> SPENCER W. KIMBALL

Image by MCCAIG

In the middle of a hot summer, and the power grid goes down. Within hours, your city descends into chaos, and everywhere you look, there is a tense feeling in the air. Without electricity, people can't keep their homes cool, their food fresh, or pump gas. You go to stock up on bread and crackers, but the supermarket shelves are bare. You watch as tempers flare and friendly neighbors turn desperate. This increasingly frequent real-life scenario is a wake-up call for anyone and highlights the urgency of being ready for emergencies. As I write this 2026 edition, Storm Amy is battering the coasts of Europe, a reminder that extreme weather is not distant news but something happening right now in our own communities.

A prepper remains calm and in control no matter what curveballs are thrown their way. When natural disasters hit, when the power grid fails, or when food is scarce, preppers are equipped and know exactly what to do.

Let me introduce myself: I'm Sam Madigan. I've made it my life's mission to help others learn important life skills, and I channel this passion by sharing what I know so that everyone can protect themselves and their families in any situation.

In my own experience, one pivotal event completely changed my outlook on life to start me on the path to self-reliance. I live in a temperate climate, so our infrastructure isn't set up for extreme weather events. Around twelve years ago, we suffered a pretty severe winter storm, and my family and I were snowed in for over a week. Anyone with young kids will appreciate the seriousness of this situation. Luckily, the ice and snow melted just before our food supplies dwindled to nothing, but it was during that week that I really grasped theimportance of never getting caught off guard again. I became laser-focused on learning everything I could about crisis readiness, and this is why I want to share this knowledge with you today.

Prepping is often misunderstood. Some see it as a paranoid obsession or about surviving the impending 'Zombie' apocalypse, but it's far from that. (Spoiler: It's less about heroics and more about the beans.)

Readiness is about responsibility, practicality, and empowerment. It's about taking control of your future, building self-reliance, and keeping you and your loved ones safe, no matter what unexpected curveballs are thrown your way. It's about living with confidence and peace of mind. And maybe a tiny bit of making sure you're the one who knows where the batteries are when the lights go out.

This book aims to provide you with up-to-date, practical, and actionable strategies for sustainable, off-grid living and emergency response planning. It covers everything from home to wilderness survival. This book is your one-stop guide to preparing for anything life throws at you, whether it's a natural disaster or just making sure you never run out of coffee. Inside, you'll find everything you need to know about:

- Food Preservation: From stockpiling to canning, you'll learn how to create a pantry that's more reliable than your local grocery store.
- Water Filtration: Water is life-literally. I'll show you how to purify and store it so you're never left high and dry.
- Home Defense and Self-Defense: You'll discover smart strategies to protect your home, your family, and yourself, because being ready isn't just about food and water.
- Medical Response Planning: From building the ultimate first aid kit to handling everyday injuries, you'll get the tools you need to keep calm and carry on in a crisis.
- Wilderness Survival Skills: Master the skills needed to evacuate safely and survive outdoors if SHTF.
- Mental Resilience: Build up your mental strength to stay calm and focused during tough times.

This book is suitable for anyone interested in emergency planning and sustainable off-grid living.It's especially great for anyone who wants to protect their kids, grandkids, grandparents, or pets.

The information in these pages is divided into different sections to navigate and find what you need easily:

At-home and shelter-in-place strategies: First, we cover essential steps for protecting yourself and your family, as staying at home is often the safest option.

Evacuation and wilderness survival: Then, we examine crucial strategies on how to safely evacuate and survive in the wilderness during an emergency.

Long-term preparedness and recovery after an emergency: This section offers valuable guidance on long-term planning and recovery, helping you and your family regain stability in the aftermath of a crisis.

So why choose this book? We include the latest information, some real-life examples, useful checklists, and how-to guides with practical advice and tips from professional survivalists as we go. This book is not just going to spew the theory; it will give you real-world, actionable, and effective strategies for you to implement immediately.

Within each chapter, you'll find a 'Resource Vault' where various books, articles, apps, commercial companies and products are mentioned or recommended, but I only do so for informational purposes. These references are solely to give you some practical tools and resources to support your own survival and readiness efforts. I do not receive any financial benefit, sponsorship, or endorsement from any of the product companies mentioned.

I encourage you to engage actively with the content as you read on. Take notes, underline key points, try out some of the tips and strategies, and make your emergency planning a priority and a continuous process. The knowledge and skills you'll gain will serve you well.

Preparedness is not just a once-off effort; it's a way of life. Nothing beats the confidence that comes from being fully equipped. So, let's do this: let's make sure that you and your family are ready for anything.

FOUNDATIONS OF PREPPING

Some years ago, a friend of mine, Bob, found himself in the middle of a wildfire evacuation, surrounded by a chaoticscene. The air was thick with smoke;neighbors were trying to rally their families and scrambling to pack essentials into their cars while their pets were panicking and running loose. Other neighbors had well-developed evacuation plans, bug out bags, and clear heads. Bob saw first-hand the stark difference between those who had a plan in place and those who didn't. As a result, he completely transformed his life by picking himself up and focusing his time and energy on learning how to be equipped to handle similar future challenges with greater clarity and confidence.

My friend's experience also gave me another gigantic push in my commitment to learning and readiness. Now it's our turn; let's get started on our own journey by taking a look at the foundational mindset that every survivalist needs.

1.1 UNDERSTANDING THE PREPPER'S MINDSET

Resilience isall about being able to overcome the unexpected.

> Sustainability is about survival. The goal of resilience is to thrive.
>
> JAMES CASCIO

A prepper's mindset follows a philosophy of self-sufficiency, resilience, adaptability, and proactive planning. At its core, survival planning means knowing how to support your family and yourself without depending on unstable outside systems that could go south at any time. Our philosophy is all about being prepared, not paranoid. It's about having the confidence and knowledge about how to tackle any situation head-on. Life is full of surprises and curveballs, but we can adapt and conquer whatever comes our way when we plan ahead.

Many people believe that survival readiness is solely about having the correct tools and supplies, but it's actually much more than that. It's about being tough when things get rough and staying calm when it counts.

We are not about reactively responding to crises after they've already happened; we want to take decisive action before things spiral completely out of control.We're all about honing our skills to stay ahead of the game, creating solid plans, and being unwaveringly set for whatever comes our way.

The prepper movement has deep historical roots, stretching back to the Great Depression and WWII, when self-sufficiency was absolutely essential for survival. Families grew their own food, preserved supplies, and made do with what they had. The Cold War era further fueled survivalism, with people building fallout shelters and participating in bomb drills. The recent pandemics, increasing natural disasters, and economic instability have really shown us the importance of being resilient. These modern influences have shaped the movement into what it is today: a practical and necessary way to make sure we and our families stay safe and secure.

Embracing a survivalist mindset demands a significant psychological shift. We live in a world where convenience is king, and most of us rely heavily on external systems to meet our needs.But just because the system we have in place works right now, doesn't mean it will work forever. We must overcome our normalcy bias: the belief that things will always remain the same. This means staying informed, building new skill sets, and being open to adapting your plans as new information becomes available.

You may have come across some misconceptions about prepping. People often think it's all about being extreme or paranoid, but that's not the case at all. In fact ,far from being paranoid, preppers are some of the most level-headed and practical people you could meet, with a deep sense of respect and responsibility. No, readiness isn't about doomsday scenarios or living in constant fear. It's actually about feeling empowered and being practical.

The psychological aspects of crisis planning cannot be overstated. A proactive, resilient mindset is a vital component of keeping ahead. Maintaining a habitual positive 'can do' outlook before and during crises helps you stay motivated and focused. This is essential to maintain a calm and collected demeanor so you can make rational decisions and keep yourself and your family safe.

We hear a lot about cultivating mental resilience, and it may seem like an empty buzzword, but stress management and mental toughness are absolutely key to navigating challenging situations.

Developing resilience, self-reliance, and independence fosters confidence because you'll know you have the skills and resources to handle emergencies. You'll also sharpen your situational awareness so you can identify what needs to be done quickly and efficiently. Overcoming common psychological barriers is a part of the prepper's journey. We need to offset our fear of the unknown by educating ourselves and taking action to stay ready. When we get stuck with inevitable information overload or when analysis paralysis takes hold, we break tasks into smaller, manageable steps so we can keep moving forward, no matter how slowly. We combat procrastination and strive to maintain high, consistent motivation, which is essential for staying on track with our contingency planning efforts.

Community and support systems play a significant role in fostering a prepper's mindset. Joining a reliable local survivalist network can be beneficial on multiple levels, providing both emotional and practical support. Building and maintaining strong community ties enhances collective resilience, too. Successful communities often share resources, knowledge, and skills, creating a robust support system for everyone.

To put it simply, understanding the survival mindset means embracing continuous learning, practicing, skill-building, self-reliance and resilience, planning ahead and adaptability, and building a community of like-minded people. This mindset forms the foundation of an effective game plan.

1.2 FAMILY EMERGENCY PREPAREDNESS PLAN

> Start by doing what is necessary, then do what's possible;
> and suddenly you are doing the impossible.
>
> FRANCIS OF ASSISI

Emergencies don't conveniently fit in with our schedule, and when they strike, panic and confusion can easily take over. But with a clear family emergency plan, when each family member understands their role and responsibilities, everyone knows what to do and where to go, reducing the chaos. For example, if a fire breaks out at night, knowing the quickest exit routes and having a designated meeting point can save lives.

It's also crucial to adapt any discussions about plans for the benefit of young children. Use simple language and include them in the practice drills by making it fun but clear so you can make sure they grasp what to do without feeling scared.Similarly, consider the needs of elderly relatives who mightrequire extra assistance or specific medical supplies.

Creating a family emergency plan involves several common-sense steps, but it takes a prepper to get the work done:

Threats: Start by outlining various types of emergencies you might face, such as natural disasters or home invasions. Your threat matrix can help you pinpoint these and consider how to adapt your planning.

Develop Templates: Include important phone numbers and addresses, family communication and check-in plans, meeting points, and evacuation routes.For instance, in the event of a flood, knowing the nearest high ground and having a route to get there can be life-saving. You'll find just such a template to download in your Free Bonuses at the end of this book.

Customize your Plans: to fit your family's unique needs; whether you have young kids, teenagers, or elderly grandparents, ensure your plan is comprehensive and covers all bases.

Supplies and Equipment: List all your supplies, including food, water, first aid kits, and tools, and customize your lists to include any special requirements. For example, a checklist for children might consist of diapers, formula, and comfort items, whereas one for elderly relatives might list medications and mobility aids.

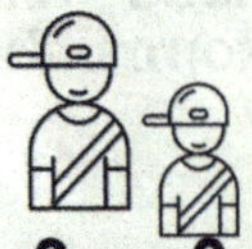

Accessibility: Ensure the checklists are accessible to all family members.

Regular Drills and Updates: Schedule family emergency drills at least twicea year. These drills help everyone remember the plan and efficiently identify areas needing improvement.

Review and update your family emergency plan regularly to reflect new risks or any changes in your family's situation. Staying informed about potential threats and incorporating this information into your plan keeps it relevant and effective. Practicing and updating the plan ensures you're always equipped to act quickly and decisively during any emergency.

1.3 FINE-TUNING YOUR EMERGENCY PREPAREDNESS PLAN

Assessing risks and understanding your family's unique vulnerabilities is the bedrock of any effective plan. To help you identify and prioritize what to get ready for, you'll need to create a thorough threat risk matrix.

Again, I know this is common sense, but you may find hazards you hadn't considered and adapt your crisis plan as a result.Also, consider that most people take a 'Yeah, yeah' approach and never get around to doing the work. So trust the process and do the work so you can build a comprehensive and successful plan

Start by evaluating personal vulnerabilities. Consider health conditions, medical needs, and family dynamics. If a family member has a chronic illness requiring daily medication, ensure you have a sufficient supply and a way to keep it cool if the power goes out. Consider how to pair people up given the layout of your home, for example.

You also need to consider your financial stability, which will affect how much you can invest in prepping supplies and emergency gear.

Geographic location matters immensely. Look at past natural disasters in your area to predict future threats. If you live in a flood-prone zone, prioritize waterproof storage solutions and evacuation routes. But also consider the curveballs that have never previously happened.

An often overlooked risk is Electromagnetic Pulse (EMP) events, which can be natural (from solar flares) or artificial (from high up nuclear detonations or electromagnetic bomb). Although they are considered low risk, EMPs can totally knock out electrical grid, leaving you without power for an extended period. The potential impact is massive, affecting everything from your refrigerator to your car's ignition system.

We need to plan for any potential human-related threats, such as crime, economic instability, civil unrest, and terrorism. Although far less frequent or likely, we also need to plan for health-related threats, such as pandemics, bioterrorism, and chemical plant accidents, which can pose significant risks. As I write this, I'm watching reports of Storm Amy battering European coastlines, while only days ago a typhoon flooded parts of the Philippines, and a 6.9 magnitude quake off Cebu rattled towns and claimed dozens of lives. Disasters strike fast and on multiple fronts. The time to prepare is always before the next one hits.

Take a look at your community's infrastructure. How reliant are you on public services like emergency medical services, law enforcement, and utility services? If these systems fail, you'll need to be self-reliant for a while.

Developing a threat matrix can really help you focus on what is important by considering possible future scenarios. Remember, you don't need a crystal ball to predict the future, just a solid plan (...and maybe an extra roll of duct tape).

List potential threats and rate them on a scale, considering factors like frequency and severity. You can source online or offline tools to create and update your threat matrix regularly and keep your response plan dynamic and relevant.

The Prepper Risk Assessment on TruePrepper ranks various threats based on their severity and likelihood. You could look at which ones are most relevant to you, but consider building a plan for all of these multiple scenarios:

Risk Level 9
- House Fire
- Home Invasion

Risk Level 8
- Cyber Attack
- Hurricane

Risk Level 7
- Tornado
- Flood

Risk Level 6
- Earthquake
- Severe Storm
- Wildfire
- Heat Wave
- Bombing/Active Shooter

Risk Level 5
- Extreme Cold
- Drought

Risk Level 4
- Pandemic
- Landslide
- Coastal Event

Risk Level 3
- Power Outage
- HAZMAT Incident

Risk Level 2
- Space Weather
- Volcano
- Biological Attack
- Nuclear Accident

Risk Level 1
- Electromagnetic Pulse (EMP)
- Nuclear Attack
- Conventional Attack
- Chemical Attack
- Radiation Emitting Device (RED/RDD)

Regular review of your threat matrix is recommended

Conduct a comprehensive inventory of your resources. Check your food supplies, water sources, and filtration systems. Assess home security measures and medical supplies. Identify gaps and surpluses. Set clear goals for your survival planning efforts. Make them specific, measurable, achievable, relevant, and time-bound (SMART). For example, aim to store a three-month supply of non-perishable food by year-end. Prioritize aspects of your crisis planning based on immediate needs and long-term sustainability. We'll cover of this later on.

You can build and execute your personalized plan by identifying actionable steps from this book and allocating timelines by breaking them down into manageable chunks. You'll need to be proactive and factor in regular reviews to adjust your emergency plan and incorporate any new information or changing circumstances. It's an ongoing process. Think of it like doing laundry: it's a bit boring, but if you don't stay on top of it, everything gets messy fast.

1.4 RESOURCE VAULT

Here's a list of additional valuable resources to help you start your journey, particularly if you are new to prepping. These books, websites, and tools offer a wealth of information on various aspects of disaster readiness, from mindset to practical skills.

If you are looking for a good overview, I recommend The Prepper's Blueprint by Tess Pennington, who provides a comprehensive guide covering everything from basic crisis response principles to advanced survival strategies.

For those interested in the historical context and evolution of the survivalist movement, Patriots: Surviving the Coming Collapse by James Wesley Rawles is an excellent book. As a former US Army Intelligence officer, Rawles brings his brand of knowledge to the table for an educational and entertaining read.

Ready. GoViz is a fantastic starting point for online resources. This website offers a wealth of information on emergency planning, including templates for creating family emergency plans, guidelines for various types of disasters, and tips for safeguarding your home. The site also provides guides tailored to different languages and setups, such as families with children, elderly family members, and pets.

You'll need an invaluable, durable, and functional multi-tool in your kit. For beginner survivalists, I recommend investing in a reliable one such as the Leatherman Wave+, which includes pliers, knives, screwdrivers, and other handy features.

Have you joined any online communities like the Survivalist Forum or Reddit's r/preppers in order to stay informed? These communities are amazing for connecting with like-minded people, getting support and inspiration, and sharing all sorts of ideas, experiences, and skills. It's a great way to learn and grow with others who share your interests.

Now we've gone over some of the basics, let's discuss what to do if you find yourself in a situation where you need to stay put for a while. We'll explore some effective strategies to help you feel secure and comfortable while sheltering in place.

> We are not preparing ourselves
> for the world welive in,
> we are preparing for the
> world we find ourselves in
>
> MICHAEL MABEE

SHELTER-IN-PLACE: HUNKERING DOWN

MEDICAL PREPAREDNESS: FIRST AID

> There's no harm in hoping for the best aslong as you're prepared for the worst.
>
> **STEPHEN KING**

Deciding whether to shelter in place or evacuate is a big decision and obviously needs careful consideration of your location's safety and structural integrity, along with your access to food, water, medical supplies, and power. If resources are limited and a family member needs special care, evacuation may be the best option. If the danger is imminent and life-threatening, evacuation may also be the best option, but ensure the evacuation routes are clear, safe, and accessible.

On the other hand, if you have adequate food, water, and medical supplies and can safely remain indoors for the duration of the emergency, if the danger outside is greater than staying home, if roads are blocked, or if authorities advise against evacuation, then you'll be sheltering in place. If you're at home, that means hunkering down and sitting it out while remaining vigilant for any updates that could change your plans.

This next part of the book is all about helping you get ready to shelter in place. We'll start with medical response planning: Creating a comprehensive first aid kit is one of the first and most important prepping tasks, in my opinion. It's a life-saving necessity, and every home should have one. And don't worry: this information will come in handy even if you end up needing to evacuate.

2.1 ASSEMBLING YOUR COMPREHENSIVE FIRST AID KIT

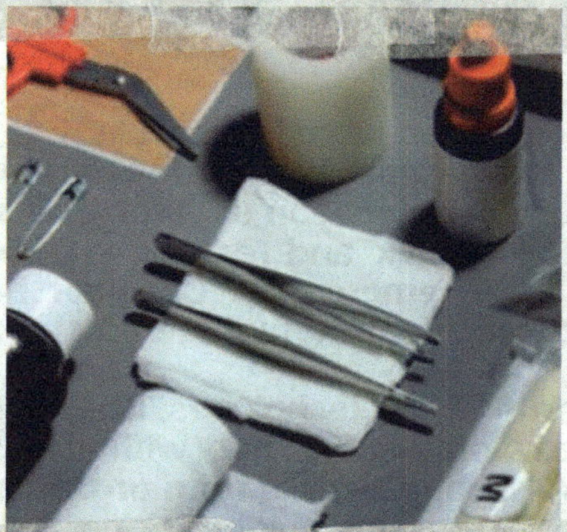

IMAGE COURTESY OF ROGER BROWN VIA PEXELS

A first aidkit is more than just a box with a few raggedy bandages inside. It's your first line of defense in an emergency and a non-negotiable must-have whether you're at home, in your car, or out in the wilderness. It's all about getting set for the unexpected and having the suppliesto turn a minor issue into something manageable instead of landing yourself in a major problem.

When assembling your first aid kit, start with the basics.

- Bandages: You'll need:
  - Bandages in various sizes and adhesive tape.
  - Gauze tape.

I recommend stocking up on cohesive bandages that are hand-tearable and stick to themselves.

- OTC Medications: Next you'll need over-the-counter medications:
  - Pain relievers like ibuprofen or acetaminophen.
  - Antihistamines for cleaning wounds.
  - Antacids for stomach issues

Medical tools are another essential component. Include:

- Scissors
- Tweezers
- Digital thermometer / Mercury thermometer
- Scalpels
- Sutures
- Emergency items like a CPR mask
- An emergency blanket
- A tourniquet

These arethe building blocks of any good kit. It's also important to customize them to fit your family's specific needs. For example, for those with special health conditions, include necessary items like asthma inhalers, EpiPens, insulin, or a blood pressure monitor. It's a good idea to create or download a medication log to track usage and expiration dates. For families with infants and children, include pediatric medications and child-sized bandages. Tailoring your first aid kit to your family's unique needs ensures that you're ready for any medical situation that might arise. You can download your free First Aid Checklist bonus at the end of this book.

Alternative and herbal remedies can also be valuable additions to your first aid kit, especially if you're living off-grid. Common herbal remedies can be really helpful, such as echinacea for colds, lavender for stress, and yarrow for toothaches. You can make echinacea tea by infusing any part of the plant in boiling water: the roots, leaves, flowers, or stems.

Growing and harvesting your own medicinal plants is an excellent way to have access to a supply of natural remedies. For example, making an herbal salve for minor cuts and scrapes using plants like calendula or comfrey can be easy and practical. To make a calendula or comfrey salve, infuse one cup of dried calendula flowers, or comfrey leaves inone cup of carrier oil (olive, coconut, or almond) by placing them in a jar for four to six weeks in sunlight or heat gently in a double boiler for two to three hours. Strain the oil, melt one ounce of beeswax, and mix it with the infused oil. You can also add a few drops of essential oils. Store in a handy tin or jar for up to a year. These remedies offer a sustainable way to manage common ailments when conventional medicine might not be available

Storing and organizing your first aid kit is just as important as what you put in it. It goes without saying that you should use clear, labeled compartments to keep everything organized and easy to find. Store your kit in a central, known location so everyone in the family can access it quickly in an emergency. Regularly check and restock supplies to ensure nothing is expired or missing in the event of an emergency. Consider having portable kits for your vehicles. That way, you'll be prepared no matter where you are.

In the next section, we'll go over some of the common first-aid techniques. Bear in mind, though, that nothing beats hands-on experience. Ideally, stay up to date with life-saving methods by taking first aid courses at least every two to three years from organizations like the Red Cross or your local community center. You may consider taking a family friendly first-aid class that will be suitable for children. You can also find plenty of helpful online resources, including instructional videos and manuals.

Ensure all family members know how to use the kit and what to do in various emergencies, and regular drills will give everyone the confidence to act quickly and correctly when it matters most.

> A first aid kit is like a parachute ...better to have one and not need it than the other way around
>
> SAM MADIGAN

2.2 FIRST AID INSTRUCTIONS

Understanding the fundamental principles of first aid is vital for any survivalist. In an emergency, the first step is to assess the situation and ensure the safety of both yourself and the injured person.

The ABCs of first aid: Airway, Breathing, Circulation These are the first things to check if a person is not moving or is unconscious. First, we assess the person's Airway and check if they are Breathing. If they are breathing, we need to check their Circulation. Also, gently move them into the recovery position to keep their airway open and prevent them from choking.

To check the airway, slightly elevate the person's chin. This will gently tilt the person's head up and back to open up the airway. If you see any obstruction, remove it.
Then, place your ear close to the person's mouth and nose to listen for breath sounds. See if you can feel their breath on your cheek, and also look down their body to check and see if their chest rises and falls.

If you think the person is not breathing normally, start cardiopulmonary resuscitation (CPR) right away.

Once they are breathing by themselves, check their pulse and look for signs like skin color and temperature. A fast pulse may suggest a fever, shock, or dehydration, while a slow pulse could indicate a problem with the heart's electrical system. If there are no signs of circulation, begin CPR immediately.

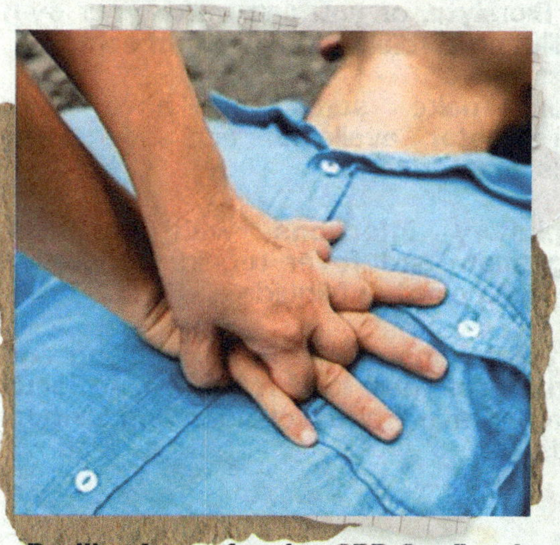

Handling serious medical emergencies requires quick thinking and action. In cases of cardiac arrest, performing CPR (Cardiopulmonary Resuscitation) can literally save lives. For adults, place the heel of your hand on the centre of the chest, then put the palm of your other hand on top and press down by 2 to 2.5 inches (5 to 6 cm) at a steady rate of 100 to 120 compressions a minute. Press down in time with the song 'Staying Alive' as a quick guide. After every 30 chest compressions, give two rescue breaths by pinching the nose, creating a seal around the mouth, and blowing for 1 second each time. Look down and check the chest rose and fell after you blew.

Position for performing CPR Cardiopulmonary Resuscitation, image courtesy of spkphotostock

For a young child, follow the advice for an adult, but only compress 2 inches deep. For infants, use two fingers in the center of the chest, just below the nipple line. Compress about 1.5 inches deep, again at a rate of 100-120 compressions per minute. After 30 compressions, give 2 light, shallow breaths; do not blow too hard as an infant's lungs have a very small capacity, and it could injure them. Cover both the infant's mouth and nose with your mouth, and give each light, shallow breath over 1 second.

Using an Automated External Defibrillator (AED) is another life-saving technique; recognize the signs of cardiac arrest and operate the AED following its instructions. It is good practice to find out where the local AEDs are kept near your location and make a note of them on your family emergency plan.

For choking, the Heimlich manoeuvre and back blows can help dislodge the obstruction. Stand behind the person, make a fist and place it just above their belly button, under the ribcage. Grasp your fist with your other hand and pull sharply upwards and inward, compressing their abdomen to expel the trapped object.

When a young child is choking and can't breathe, stand behind them, place a fist just above their belly button, and give quick, upward thrusts until the object is expelled. If an infant is choking, hold them face down along your forearm, head lower than their body, and give 5 firm back blows between the shoulder blades. If that doesn't work, turn them face up and give 5 chest thrusts using two fingers in the center of their chest. Repeat back blows and chest thrusts until the object is cleared or they become unconscious, then begin CPR if needed.

In case of serious bleeding, act quickly by applying pressure on the wound and use a tourniquet to limbs if needed. If the person is in shock, lie them down and keep them calm and warm. Continue to watch out for rapid breathing, a weak pulse, or any confusion. Elevate their feet by about 12 inches unless you suspect their head, neck, or back is injured or if you think they have broken leg or hip bones.

Treating common injuries and illnesses requires a systematic approach:

For cuts and lacerations, clean the wound with an antiseptic, disinfect thoroughly, and then dress and bandage it to prevent infection.

Strains and sprains can be managed using the RICE method-Rest, Ice, Compression, and Elevation.

Signs of fractures include swelling, bruising, and inability to move the affected limb. Immobilize the injured area and create a splint to keep the bone steady.

Burns vary in severity. Minor burns can be treated with cool or room temperature water (notice). Aloe vera soothes the skin, and over-the-counter pain relievers manage pain.

Animal bites, puncture wounds, and lacerations should be thoroughly cleaned, and medical advice should be sought if possible.

We also need to spot or recognize illnesses that may not be obvious so we can treat them:

Dehydration can be serious, so keep an eye out for signs like dizziness, dry mouth, or dark-colored urine, and rehydrate with water or an electrolyte solution. You could use Dioralyte, or you can make your own by diluting cordial with a pinch of salt.

Fever: make sure you have your thermometer available, and use fever meds as needed.

Respiratory infections with coughing, congestion, and fever require supportive care, with plenty of liquids and rest.

Gastrointestinal issues with vomiting and diarrhea can lead to dehydration, so focus on rehydrating and consider antidiarrheal meds.

Image courtesy of Lauzla

Remaining calm and composed is essential. Panic can make the situation worse, so take a deep breath and consciously shift your focus to the steps you need to take. Always perform a primary survey by quickly checking for life-threatening conditions, starting with the ABCs.

I'm guessing you don't want to find yourself in the middle of a crisis, suffering from a serious illness, right? So, we need to stick to our usual basic preventative health measures, such as regular handwashing and basic sanitation, particularly when everyone is under stress and immune systems are depressed. Vitamin supplements are even more important during these stressful times to stay as fit and healthy as possible.

When it comes to emergencies, keeping clean can be tough, especially if you have to leave in a hurry. Don't forget to pack essentials like hand sanitizers and baby wipes. If you don't have them, try to make do with what you have.

You can dig a hole to use as a toilet, away from where you are eating and sleeping. You could even create something to siton by using a toilet seat on a box witha hole in it. You can move everything to a freshly dug hole when the old one is full. Consider building a more complex compost toilet to reduce odors. This works by separating out solids and liquids and adding sawdust or cat litter to the solids, for example. More details on this will be provided in the Wilderness Survival section.

2.3 ASSEMBLING YOUR PET EMERGENCY KIT

You can dig a hole to use as a toilet, away from where you are eating and sleeping. You could even create something to siton by using a toilet seat on a box witha hole in it. You can move everything to a freshly dug hole when the old one is full. Consider building a more complex compost toilet to reduce odors. This works by separating out solids and liquids and adding sawdust or cat litter to the solids, for example. More details on this will be provided in the Wilderness Survival section.

A pet emergency kit must cover many bases. It seems clear that we need to start with non-perishable pet food and an ample water supply. For smaller pets like small dogs and cats, plan for 1 to 1.5 gallons of weekly water allocation per pet. For larger pets, estimate 1.5 to 3 gallons per week.

Don't forget to pack all medications your pet needs, along with a copy of their vet records and recent photos, in case they get lost. Also, include basic first aid supplies for pets, like:

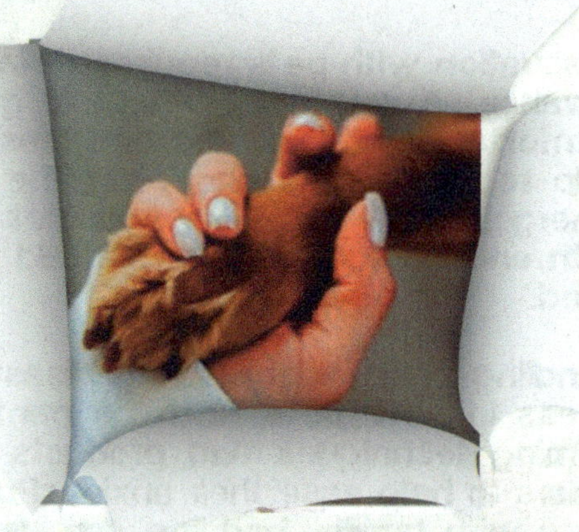

Image courtesy of Ivan Babydov

- Cohesive bandages (easily tearable and stick to themselves).
- A good antiseptic scrub like Hibiscrub (to wash out potential wounds). And tweezers (for minor injuries).
- A pet-specific first-aid manual can help guide you through many basic treatments.

Comfort items are just as important:

- Pets can become stressed in emergencies, so having familiar items like favorite toys, blankets, or bedding can provide a sense of normalcy. For instance, including a favorite toy or stuffed animal can offer much-needed comfort to a stressed pet.
- Use one of their blankets to cover the top of their crate or carrier, particularly when evacuating. This will obscure their view of what is happening and can help them feel more protected, safe, and calm.
- Calming aids, such as pheromone sprays or calming collars, can also be beneficial.
- These small steps can make a significant difference in how your pet copes with an unfamiliar situation.
- Bear in mind, different types of pets have different needs. For example, dogs and cats require specific considerations:
- For dogs, ensure you have a sturdy leash and harness at the ready.
- For cats, a secure carrier and some litter are essential, so you'll need to stock up.
- Small animals and birds have unique requirements:
- Extra bedding material is needed for small mammals.
- A travel cage for birds can make evacuation a lot easier.
- Customizing the kit for your specific pets ensures that all their needs are met, regardless of the emergency scenario.

2.4 TIPS FOR KEEPING PETS COMFORTABLE

When disaster strikes and you need to shelter in place, routines often go out the window, which can be very disruptive for pets. Creating a safe and secure indoor environment is important to help them stay calm despite the unfamiliar new circumstances.

Choose an interior room away from windows and exterior walls to minimize exposure to hazards. You'll have gathered an adequate supply of food, water, and hygiene items for your pets from your checklist.

Keeping their feeding and exercise schedules as consistent as possible will help to maintain a sense of normalcy. This stability can significantly reduce their stress levels during chaotic times.

Evacuation with pets requires detailed planning. Identify pet-friendly shelters and hotels in advance so you know where to go if you need to leave home. Make sure your pets are comfortable with carriers and leashes by practicing evacuation drills with them. This will help reduce stress and offer the best chance for them to cooperate during an actual emergency. By following your pet evacuation readiness checklist, you'll be well prepared with enough food, water, first aid kit items, any medications, and comfort items they will need.

Handling pet anxiety during disasters is another challenge. Learn to tune into signs of stress in your pets, such as panting, pacing, or hiding. Offer physical comfort and use calming techniques and products like soothing music, pheromone diffusers, or calming collars to help ease their anxiety.Providing a safe space where they can see you, such as a crate with familiar bedding, can also be comforting. Monitor their behavior and offer plenty of reassurance, as this can make a significant difference in how they cope with the situation.

Gradually reintroduce them to a new environment, for example keep them in a small safe space first to acclimate before allowing them into larger spaces. Reestablishing routines will support their adapting and recovery.

Post-disaster pet care involves thorough check-ups and continued monitoring. After the immediate danger has passed, use common sense and check your pets for injuries. Monitor their behavior for any signs of trauma. Look for cuts, bruises, or signs of stress and seek veterinary care if needed or possible.

2.5 RESOURCE VAULT

If you wish to expand your knowledge of medical preparedness and first aid, here are some additional resources with practical advice to enhance your readiness.

The Survival Medicine Handbook by Joseph Alton, MD, and Amy Alton, ARNP, is a great addition to any prepper's bookshelf. It covers a wide range of common medical scenarios for when professional help is unavailable.

For those interested in natural remedies, James Green's The Herbal Medicine-Maker's Handbook is an excellent resource. It provides detailed instructions on creating herbal remedies, from tinctures to salves. This book is especially useful for off-grid living, where access to conventional medicine might be limited.

Online guides such as the American Red Cross website (redcross.org) offer a wealth of information on first aid training and emergency response, with instructional videos and manuals, and you can sign up for first aid courses.

Investing in a high-quality first aid kit like the Adventure Medical Kits is a wise choice. These kits are designed for a variety of scenarios and include a wide range of medical supplies, ensuring you're well-prepared for any situation.

A reliable digital thermometer, such as the Braun ThermoScan, is an essential tool that you likely already have in your home. A mercury thermometer is also handy as it won't require batteries.

These tools are super helpful but remember; the key is to keep learning and practicing too so you'll feel confident no matter what.

We've covered the basics of first aid, a vital necessity in your skill set, and caring for pets during an emergency. Next, we'll explore strategies for food stockpiling and make sure you have the supplies to sustain you and your family through any crisis.

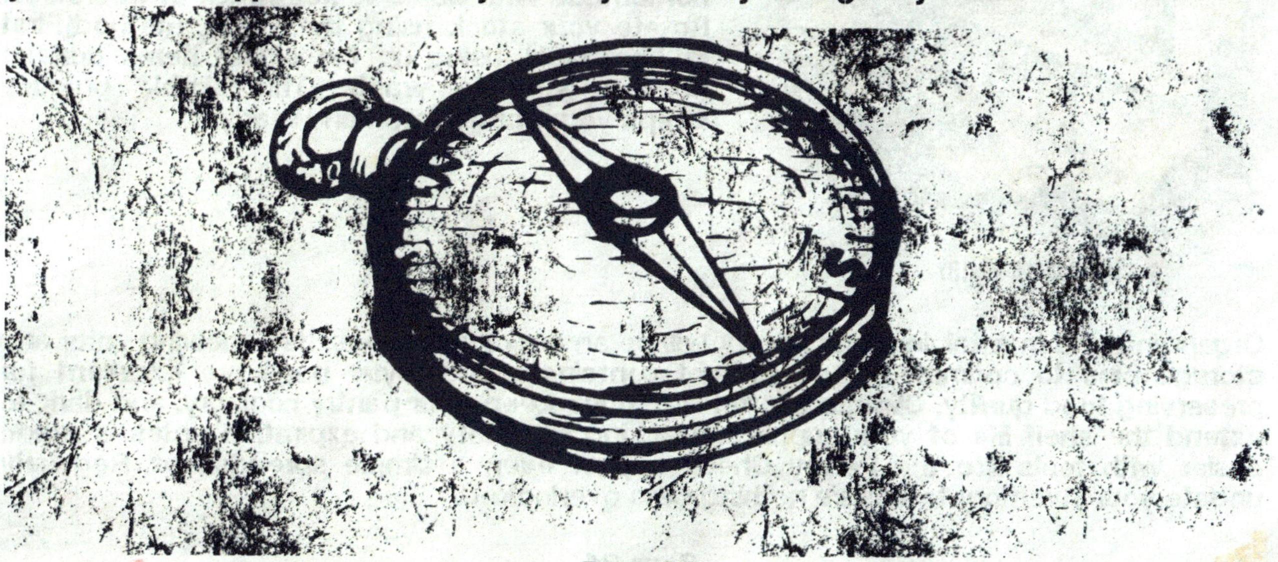

FOOD PREPAREDNESS

The first time I saw empty grocery store shelves, it was a shock. Early one Spring a few years back, bad weather hit our city. During a storm, which was nothing catastrophic but was enough to cause a rush on essentials, stores were literally emptied within hours.The sight of people scrambling for the last loaf of bread or carton of milk was a sobering reminder of exactly how fragile our supply chains really are. In late 2025, as Spain battled its worst wildfire season in decades and South Korea faced simultaneous blazes, communities also saw their supply chains collapse almost overnight. We can't count on supply chains and stores; there must be another better way.

3.1 BUILDING YOUR PREPPER PANTRY

A homesteader's pantry is a dedicated space where you stock up on extra food and necessities, almost like your own personal grocery store only better, providing you with enough supplies to last through emergencies. No more scrambling for that last toilet paper roll in the store! Consider it your 'edible insurance policy'. No paperwork required, just guaranteed food security and peace of mind when you really need it. Plus, you'll be in a position to help others in times of real emergency. A well-organized pantry can be the difference between stress and stability when SHTF.

Sustainability and self-sufficiency are the backbone of a homesteader's pantry, giving you control over your food supply. When you take charge and have the foresight to prepare for disruptions, whether from natural disasters, economic instability, or supply chain issues, you can rely on your own resources without constant trips to the store.

A WELL-STOCKED PANTRY

Inventory

Start by carrying out an inventory and organizing your pantry for maximum efficiency. Categorize food supplies into grains, proteins, canned goods, and other essentials. Unless you want to risk playing 'refrigerator roulette', label everything homemade with dates to keeptrack of freshness. Rotate your stock using the F.I.F.O. method(First In, First Out), using up the oldest items first so nothing goes to waste. This simple practice keeps your pantry fresh and ready.

Organizing and maintaining your pantry will be an ongoing process. Use shelving units and storage bins to optimize space. Airtight containers and Mylar bags are excellent for preserving food quality. Climate control is a must; keep your pantry cool, dry, and dark to extend the shelf life of your supplies. Tracking inventory and expiration dates is made easier with tools like the Pantry Check app or even a simple spreadsheet. Regularly update your inventory to ensure nothing gets overlooked.

Organizing and maintaining your pantry will be an ongoing process. Use shelving units and storage bins to optimize space. Airtight containers and Mylar bags are excellent for preserving food quality. Climate control is a must; keep your pantry cool, dry, and dark to extend the shelf life of your supplies. Tracking inventory and expiration dates is made easier with tools like the Pantry Check app or even a simple spreadsheet. Regularly update your inventory to ensure nothing gets overlooked.

Budgeting and sourcing can be challenging in today's economy, but there are strategies to keep costs down. Make your budget stretch further by purchasing items like rice, beans, and pasta in bulk. Sprouts Farmers Market and Thrive Market are popular choices among homesteaders. If you have one, try your local farmer's market. Set a monthly budget for pantry building, separate from other expenses. Use apps like YNAB (You Need A Budget) to track expenses and stay on course.

Sustainable practices are pivotal for long-term success. Growing and preserving your own food reduces reliance on external sources. Renewable energy sources like solar dehydrators should ideally be incorporated for food preservation. We will talk about this later on in this chapter.

Building or joining a community network offers a ready-made support system for pooling knowledge and resources together. Strengthening your survival skills with this kind of collaborative approach also forges a sense of camaraderie and community.

Focusing on these principles will help you create a robust prepper pantry and ensure your family's well-being when sheltering in place at home.

Here's a handy table you can refer to with the shelf life of various foods:

Food item	Shelf life
white rice	indefinately if stored correction
brown rice	6 months
pasta	2 years
dired beans	1+ years
coffee	2-20 years
olive oil	2 years
chocolate (dark)	up to 2 years

Food item	shelf life
Canned goods (low acidity)	2-5 years
canned goods	high acidity
dried herbs and spices	1 year
dired fruits	1 year
tea	6-12 months
nuts and seeds	6 months
baking supplies	approx 6 months

TABLE 1: SHELF LIFE OF VARIOUS FOOD ITEMS. BASED ON GENERAL PREPPER GUIDELINES.

3.2 CANNING: STEP-BY-STEP INSTRUCTIONS

Canning is a handy, time-tested method for preserving food. It ensures you have nutritious supplies on hand even when fresh produce is scarce. The process involves placing food in jars and then heating them to a high enough temperature to destroy any microorganisms and inactivate enzymes, preventing spoilage and contamination. I think canning is a bit like putting food in a time capsule, except instead of curious archaeologists, future-you gets to open it when things go sideways.

There are two main types of canning:water bath canning and pressure canning. Water bath canning is suitable for high-acid foods like pickles or fruits, whereas pressure canning is required for low-acid foods such as meats, broths, and vegetables. Safety is a top priority in canning to avoid food borne illnesses, so follow all the proper procedures

A pressure canner is crucial for low-acid foods as it reaches the higher temperatures needed to kill harmful bacteria like Clostridium botulinum. Pressure canners can also be used for high-acid foods, but water bath canners, while simpler to use, are only suitable for high-acid foods. Both types of canners can be commonly found at well-known online stores or specialty kitchen stores. DIY versions can be made using large stockpots and racks, but you'll have to ensure they meet safety standards for canning.

To get started, you'll need some essential equipment. A good set of canning jars is a crucial component of the whole process. Choose jars made for canning, such as canning-specific Mason jars. They should be durable and made to withstand high temperatures. A headspace measuring tool ensures you leave the correct amount of space between the food and the jar lid, which is vital for a proper seal. Jar rings and lids are also necessary; rings can be reused, but lids should be new each time to ensure a tight seal. You'll need jar racks to keep jars from touching the bottom of the canner, jar lifters to handle hot jars safely, and a ladle and funnels for transferring food easily into the jars. You can prevent spoilage by removing air bubbles with wooden chopsticks or other bubble removers..

IMAGE COURTESY OF CONNIE T BALLASH

Essential Canning Prep List:

- Canning jars (such as Mason jars designed for canning)
- Headspace measuring tool
- Jar rings (reusable)
- Jar lids (non-reusable)
- Jar rack
- Jar lifters
- Ladle
- Funnels
- Chopsticks or bubble removers

Water Bath Canning

Canning requires a series of meticulous steps. Here is a summary of the steps involved:

- Get the food ready: wash, peel, and cut as needed.
- You don't need to sterilize your lids and jars if your canning process will take longer than 10 minutes. Otherwise, boil them for at least 10 minutes to sterilize them before starting the canning process.
- Fill the jars with high-acid foods like fruits or pickles, add hot water (not boiling), leaving the recommended headspace.
- Wipe clean all the jar rims before placing the lids on and securing them with the rings.
- When water bath canning, cover the jars with water (or follow your pressure canner's manufacturer's instructions to process your jars). As a general rule, process for 10 minutes if you are using four-, eight-, or twelve-ounce jars and for 15 minutes if you are using sixteen-ounce jars. Make sure to let everything cool for 10 minutes prior to removing the jars from the pot or water canner.
- Once processed, check for a proper seal by pressing the center of the lid; it shouldn't flex up and down. Water bath canning is really straightforward as long as you follow the steps. This method is perfect for preserving seasonal fruits and making jams and spreads.

Here's an easy recipe for homemade strawberry jam:

- Cook strawberries with sugar and lemon juice until thickened.
- Then fill jars, leaving a 1/4 inch headspace.
- Process in a water bath canner for about 10 minutes, adjusting for altitude if necessary by boiling for longer if you are at a higher altitude, 1000 to 3000 feet, boil for an extra 5 minutes. From 3000 to 6000 feet, boil for an extra 10 minutes and for altitudes above 6000 feet, boil for an extra 15 minutes

Pressure Canning

Pressure canners use pressurized steam to produce the required temperature of 240 degrees Fahrenheit (115 degrees Celsius) or higher. This temperature is needed to destroy the bacterial spores that naturally occur in these foods. A vacuum is formed as the jars cool, sealing the food inside the jars and preventing any new microorganisms from getting in.

Pressure canning requires a bit more precision and care than the water bath process. We need to carefully operate the canner at a pressure that accounts for changes in altitude because higher elevations lower the boiling point of water. So, a higher altitude requires increased pressure to process and preserve food safely. You'll need to watch the pressure during the process, too, to ensure it stays constant.

IMAGE COURTESY OF STEPHEN WHITE

As a general rule of thumb, weighted gauge models should be operated at 10 pounds of pressure at altitudes below 1,000 feet.

However, at altitudes above 1,000 feet, the pressure should be increased to 15 pounds. Dial gauge canners should be used at 11 pounds of pressure for altitudes between 500 and 2,000 feet. Consult your manufacturer's recommendations, as safety is key when using pressure cookers.

Michigan State University Extension recommends the following steps to ensure proper pressure canning:

- Put two to three inches of hot water into the pressure canner.Put the filled jars on the rack inside the canner, leave the recommended headspace, and close the lid to 'fingertip' tight. If a jar is shut too tightly, air cannot escape during processing, which could buckle the lid and break the seal. A good rule of thumb is to place the band on the jar, turn it until you feel a little resistance, then continue turning it an extra quarter-turn more.
- Leave the vent port off or open the petcock. Turn up the heat to the highest setting and heat the canner until you see steam flowing freely from the vent port or petcock.
- Let the steam escape steadily for 10 minutes to drive the air out of the canner.
- Place the weight on the vent port or close the petcock, and the canner will raise to the proper pressure over the next 3 to 5 minutes.
- Start your processing time countdown once the dial gauge or weighted gauge reaches the recommended pressure(refer to the above recommendations and the manufacturer's directions).
- Regulate the heat to maintain steady pressure. If the reading decreases from the recommended pressure, bring the pressure back up and start the timing process again.
- Once the proper processing time is completed, turn off the heat and watch until the dial returns to zero on a dial gauge canner, or wait 30 to 45 minutes for a weighted gauge canner (see manufacturer's directions). Wait an extra 10 minutes before opening the lid.
- Use your jar lifter to remove the jars and place them carefully on a towel or cooling rack.

- Leave the jars resting on your counter for 12 to 24 hours. Do not disturb them.
- Once the contents have cooled down, make sure to check that the lids have sealed properly.
- If any of the jars did not seal, you can reprocess the contents again according to the recipe directions. Otherwise, the jar must be refrigerated and eaten within one week. Alternatively, after readjusting the headspace to one inch, the jar can be frozen and stored for up to one year.
- It is recommended that dial gauge pressure canners be calibrated each year for safe food preservation. Pressure canners with weighted gauges do not need to be tested.
- As always, label and date your jars and store them in a cool, dark place. High-acidity canned foods can last up to 18 months, while low-acidity canned foods can last 2 to 5 years.

Basic Recipe for Canning Chicken Soup:

- Gather your ingredients: cooked chicken, vegetables, and broth.
- Fill the jars, leaving one inch of headspace, and remove all air bubbles.
- Seal the jars as indicated above and place them in the pressure canner, ensuring they are not touching.
- Process at the recommended pressure and time based on your altitude.
- This method ensures your low-acid foods are safe to store and eat for up to two years or even up to as long as five years later.

Canning seasonal vegetables, like tomatoes and green beans, offers a way to enjoy your garden's produce year-round. For tomatoes, blanch, peel, and quarter them before filling jars and processing. Green beans can be raw-packed or blanched before canning. Achieving the best flavor and texture comes with practice, but following recipes and guidelines helps.

Safety is always a primary concern in canning. Regularly check seals and consume any jars with broken seals immediately. Recognize signs of spoilage, such as bulging lids, leaks, or off-odors, and discard any questionable jars.

3.3 DEHYDRATING FOOD: TECHNIQUES AND TIPS

Dehydrating food is an easy, practical method for long-term storage. It preserves the nutrients and flavors, giving you lightweight, space-saving options. Dehydrated foods retain much of their nutritional value and can be easilyrehydrated when needed, making them ideal for both everyday use and emergencies.

IMAGE COURTESY OF SHIRONOSOV

Electric Dehydrator

The equipment you choose for dehydration can vary. Electric dehydrators are popular for their convenience and efficiency. Models like the Nesco Snack master or the Excalibur Dehydrator are widely recommended. They offer adjustable temperature settings and multiple trays for drying various foods simultaneously. Using an oven for dehydration is another option.

Set your oven to a low temperature, around 140°F, and use baking sheets to spread out the food. Prop the oven door open to allow moisture to escape and leave to dehydrate for between six and twelve hours depending on the item.

DIY Solar Dehydrator

Solar dehydrators are a sustainable option, especially for those living off-grid. They can be handy for dehydrating small slices of fruits, meats, or garden herbs. DIY solar dehydrators can be constructed using simple materials like wood, mesh, and glass.
Plans are available online, such as those from Mother Earth News, which detail how to build a solar dehydrator that can dry 10 pounds of food in two sunny days.

Here is one way to re-purpose an old table and create a your own solar dehydrator from Gardening Australia:

- Convert a table into a dehydrator. It's already flat, which gives you a good base to work from.
- Reuse some second-hand timber offcuts to use as walls around the table.Cut to size and use galvanized screws to secure them. Using a clamp will help to keep the offcuts as level as possible while you screw them together to keep everything squared up.
- At this point, your structure will look like a large tray on legs. Next, get some corrugated iron, cut it to size to fit on the table within the frame, and paint it black. When dry, put it on the table, which is now the base of your solar dehydrator. Don't secure it to the table so you can lift it out for cleaning later. The corrugated iron will absorb the heat and re-radiate it up through the fruit and vegetables you are drying.
- You will need to drill a few holes along one of the short sides of the frame for ventilation and staple some fly screen or netting over the holes to keep bugs out. If the staples don't fully embed into the timber off cut, hammer them down to buckle them over for safety.
- Next, you'll want to create an elevated shelf within the frame. Use some old architraves to place on the inside of the frame to make the raised part that you can rest your netting shelves on.You'll be spreading out the food you are dehydrating on these shelves.
- The shelf itself can be created by making some smaller frames (you may need to place several side-by-side inside the table frame). These smaller frames should fit inside and rest neatly on the architraves. You'll be able to move them in and out as needed.
- Cover these smaller frames with mesh netting to rest your fruit on, and place the now covered frames on the architraves to make your elevated shelf.
- An ideal cover is a piece of glass, maybe from an old window the same size as the table with the frame. You can rest the glass on top of the frame, but be careful when applying fixings like hinges to the frame and the wood.
- A helpful tip: before placing the glass on the table, it's a good idea to pre-drill the holes in the window frame around the glass beforehand. This helps prevent the glass from splitting. Check you have a tight seal between the wooden table frame and the window frame for the glass panel, and add a rubber or brush seal for extra protection to prevent insects from getting in and spoiling the food.
- To make it mobile, you can add wheels to the bottom of the side with ventilation holes. On the opposite side of the table to the ventilation holes, cut both legs short to the same height so the dehydrator will sit at an angle and absorb the sunshine and the heat will rise towards the ventilation holes you drilled and allow moisture out. You'll need to prop up the table from underneath so you can comfortably cut the legs.
- Use your dehydrator on sunny days to take advantage of all that free solar energy and dry your fruit.

Quick Pro Tips

- Fruit with lower moisture content will dry more quickly.
- You should be able to comfortably dehydrate a couple of pounds of fruit in one day with your solar dehydrator.
- The temperature inside the dehydrator should ideally reach between 122°F to 140°F.

If you don't want to make your own one, the Sun Oven isa popular model is a sustainable option that is easy to use.

Food Preparation

When dehydrating fruits, vegetables, and herbs, the preparation is key. Make sure you have even slices, and blanch any vegetables if you want to preserve the color and texture.

When you're using an electric dehydrator, just set it to the recommended temperature, usually between 125°F and 135°F, and then lay out your slices on the trays. Remember, drying times can vary. The drier the food is to start with, the less time it will take to dehydrate. For example, apples take around 6-12 hours, while tomatoes might take up to 10 hours. And don't forget to store your dehydrated foods in airtight containers to keep them nice and fresh.

A simple method for dehydrating apple chips involves:

- Slicing apples thinly.
- Dipping them in lemon juice to prevent browning.
- Drying them at 135°F until they are crisp.
- Meats and Vegetable Protein

Dehydrating meats and proteins requires specific considerations:

- Fatty meats are not suitable for dehydration because they can spoil quickly. Leaner cuts are ideal and make a nice jerky, the superhero of snacks-ready to save the day when hunger strikes.
- Making beef jerky involves slicing lean beef thinly, marinating it in a mixture of soy sauce, Worcestershire sauce, and spices, and then drying it at 160°F until it reaches the desired texture.
- Chicken can be dehydrated by cooking it first, shredding it, and then drying it at 145°F.
- Fish can also be dehydrated. Thoroughly dry the fish with a towel to remove excess oil before slicing and placing it in the dehydrator. You could marinate salmon in soy sauce with brown sugar, minced garlic, and lemon juice or add sesame oil, soy sauce, and ginger to sushi-grade tuna. Then, set your fish to dehydrate at 145°F for 5 to 7 hours.It will keep in the refrigerator for one month or the freezer for three months.
- You can dehydrate crab and shrimp, for example, using a similar process. When bent, fully dried shrimp will snap.
- It is also possible to dehydrate tofu by slicing and pressing it to remove moisture and then dehydrating it at around 135°F for around 6 to 8 hours.

Rehydrating Dried Foods

Rehydrating dried food is pretty straightforward. Most of the time, you just soak it in water or broth for anywhere from 15 minutes to a few hours, depending on the food. Boiling water can speed things up, especially for tougher foods like dried meat.

But if you're making a stew or soup, you don't always have to rehydrate first. For small cubes or shredded dried ingredients, you can toss them straight into the pot. The liquid in the dish will do the work for you while everything simmers, blending the flavors and saving you time.

For more detailed guidelines on dehydrating various foods, check out sites like the Backpacking Chef, which provides plenty of step-by-step instructions and tips.

3.4 FERMENTATION TECHNIQUES

Fermentation is an ancient method that not only keeps food fresh but also boosts its nutritional value and taste. Veggies like carrots, cabbage, and cucumbers, plus dairy and grains, are perfect for fermenting. The process involves giving your veggies a good wash, chopping them up, and adding some salt to draw out the moisture and create a brine. The salting step is critical as it impedes the growth of harmful bacteria while promoting beneficial bacteria. Then, just pour off the excess liquid.

Traditional crocks, glass jars, and food-grade plastic containers all work well as fermentation vessels. Each type of container has its uses. For instance, crocks are ideal for large batches, while glass jars are perfect for smaller quantities. Ensure your containers are clean and free from cracks. A good seal is vital to keep out air and contaminants.

The fermentation process is straightforward but requires attention to detail. Pack your prepared vegetables tightly into the container, and ensure they are fully submerged in the brine by adding (non-chlorinated) water. It is important to keep the vegetables submerged using a fermentation weight or a smaller jar filled with water. Cover the container with a cloth or a loosely fitting lid to allow gases to escape. Keep the container at room temperature, ideally between 60-70°F, and check it daily, and start tasting after a few days. Watch for bubbles, a tangy smell, and a change in taste to track progress.

Remember, proper storage is crucial to make your fermented foods last longer.Once they have fermented to your liking, move them to a cooler spot like a refrigerator or root cellar. This will slow down the fermentation process and lock in their delicious flavors. With the right storage, your fermented foods will last for several months, providing you with a steady supply of nutritious meal enhancers.

Fermented foods like sauerkraut, kimchi, kombucha, and kefir can add a kick of flavor to your meals and are great for your gut health and digestion.

If you're adventurous, you can even experiment with fermenting grains to make a sourdough starter for baking bread. Mix equal parts of flour and water in a clean jar (about ¼cup each).Leave the mixture loosely covered to air at room temperature for about 24 hours. Every day for about a week, remove half of the mixture and feed it with equal amounts of fresh flour and water. Over time, natural yeasts and bacteria will ferment the grains. Once your starter is active and bubbly and has a tangy smell, the sourdough starter is ready for use in baking.

3.5 CURING MEAT

An ancient technique that can transform your excess protein into enduring, tasty staples is curing meat. Meats like pork, beef, chicken, fish, and game are good options for curing. Although less prevalent, poultry can be used to make delectable goods like prosciutto from duck. Fish and game provide a wide range of alternatives, with products like jerky made from venison or smoked salmon having deep flavors and long shelf lives.

You'll need to understand curing methods to make the most of them. You can check out some of the resources at the end of this chapter to explore this technique more, but here's a quick taster so you can decide if curing is for you.

Dry curing involves rubbing meat with a mixture of sugar, salt, and spices and then allowing it to cure in a controlled environment. This is a great way to add strong flavor to meats, just like prosciutto or pancetta. During wet curing, also known as brining, meat is soaked in a saltwater solution, which can give it added taste and moisture, but bear in mind that this method won't really lengthen the meat's shelf life. Using the proper ratio of curing salts to assure safety and flavor balance is one of the fundamentals of salt and sugar curing. Meats are further preserved and flavour-enhanced by smoking. As we'llsee in the next section, you can employ hot smoking for meats that are richer and more robust or cold smoking for delicate foods like fish.

Simple substances and equipment are needed for curing. Nitrates and curing salts inhibit the growth of microorganisms and improve flavor. It's really important to use the right type and amount.

You'll need equipment such as a curing chamber, smoker, and storage containers. Modern equipment like digital thermometers and hygrometers can help maintain the ideal curing environment. If you're looking to save some money, you can try DIY options like turning an old fridge into a curing chamber, or you can easily transform any metal drum into a simple smoker. Use these tips and your common sense to ensure your curing process is safe and consistent.

The Dry Curing Process

To start your dry-curing process, choose a lean cut of meat to clean and trim, then rub the meat with your salt and spice mixture. Time and temperature are critical; most meats cure best between 50 and 60°F. The curing time varies-pork belly for bacon might take a week, while a whole ham could take several months. You'll need to keep the temperature steady, as fluctuations can spoil the meat.

When you're curing meat, keeping it safe from botulism and other food borne illnesses is super important:

- Make sure you use curing salts correctly and store your meat in a cool, dry place with good air circulation.
- Keep in mind that different cured meats have different storage times-some can last for months, while others only last for weeks.
- When it comes to food safety, it's important to be on the lookout for signs of spoilage in meat, such as off-odors, discoloration, and slimy textures. If you're not sure, it's best to play it safe and dispose of it to reduce the risk of food borne illnesses.

- Store it in a low-humidity environment of around 60%-70% at a temperature between 50 and 60°F.
- You can also wrap it in butcher paper or use breathable containers that will limit the build-up of moisture.
- And if you are living off-grid with a naturally cool cellar,that's the perfect spot to cure your meat.
- Once the curing process is complete, vacuum seal your meat to keep it safe from air and moisture to preserve it even longer.

If you need to handle large quantities, you can scale up by either building larger curing chambers or using shared facilities in a prepping community.

3.6 SMOKING MEAT

IMAGE COURTESY OF PROPHOTO

This method of preserving meat has been used for thousands of years. It involves a smidge of art, a pinch of science, and 100% making your neighbors jealous every time you get smoking. This method is perfect for smoking salmon and jerky, creating that sought-after rich, smoky taste that everyone loves.

When you hot smoke meat, you're essentially cooking it while infusing it with that delicious smoky flavor at temperatures between 165°F and 250°F. It's perfect for larger cuts like brisket or ribs. The type of wood you use affects both the taste and safety of the meat. For instance, hickory, apple, and cherry woods add unique flavors, while avoiding resinous woods like pine helps prevent a bitter taste and the absorption of harmful compounds.

First, you want to cure the meat with a mix of salt, sugar, and spices to draw out moisture and add flavor. Trim off any excess fat so the meat smokes evenly. When it comes to flavor, seasoning is key: for example, try a simple mix of salt, pepper, garlic, and paprika for that extra kick. Before smoking, let the meat rest and soak up the flavors for a few hours or overnight in the fridge. All of this ensures that the meat is flavorful and at its best for the smoking process.

When it comes to smoking equipment, you have several options. Electric smokers are easy to use and very convenient and maintain a consistent temperature with minimal effort on your part. But please keep in mind the following information:

There are various advantages to using different types of smokers. For example, charcoal smokers like the Weber Smokey Mountain provide a classic smoky flavor and are appreciated for their reliability by many preppers. Propane smokers offer a good balance of taste and convenience as they heat up quickly and maintain steady temperatures. DIY smokers, such as those made from metal drums or old refrigerators, can be effective and economical for those who are looking for hands-on projects.

When you're smoking meat, make sure to preheat your smoker to the desired temperature before you start.Then, place the meat on the smoker grates, making sure there's enough space for the smoke to circulate around the meat.

This helps to ensure that the meat gets that delicious smoky flavor evenly. Use a reliable meat thermometer to monitor the internal temperature of the meat throughout the smoking process. For hot smoking, aim for an internal temperature of 165°F for poultry and 195°F for larger cuts like brisket. Baste the meat occasionally, and add wood chips or chunks as required to keep a steady smoke going. Check for doneness by ensuring the meat has reached the appropriate internal temperature and has a desirable texture.

Make sure to store your smoked meat properly after it's done. When cooled, wrap it and hermetically seal it to maintain its freshness. As a guide, beef will keep in the refrigerator for up to a week and in the freezer for months. Don't forget to make sure to clean surfaces and utensils properly to prevent cross-contamination and control the humidity to keep mold from forming. Your smoked meat will stay delicious and safe to consume as long as it is well-stored.

Smoking meat off-grid can be a really rewarding experience. Techniques for smoking without electricity include using a charcoal smoker or building a smoker from available materials. Scale the process by smoking larger quantities at once and then storing the meat in a cool, dry place. You could consider creating a dedicated space to regularly smoke meat by building your own smokehouse.

3.7 FREEZE DRYING

Freeze-drying involves removing moisture from food while retaining its nutrients, flavor, and texture. The process freezes the food and then applies a vacuum to remove the water content. There are huge benefits and some cons to freeze-drying food: the benefits include long shelf life (up to 25 years in some cases), lightweight storage, and easy rehydration. Home freeze-dried food is tastier than store-bought freeze-dried meals. It is particularly useful for preserving fruits, vegetables, meats, and complete meals, making it ideal to stock your homesteader's pantry with. However, the cons are that it can be expensive, with most freeze-drying machines for home use costing several thousand dollars, but this may pay for itself over time if you use it frequently. Store-bought freeze-dried food is also more expensive compared to other preservation methods like dehydrating or canning.

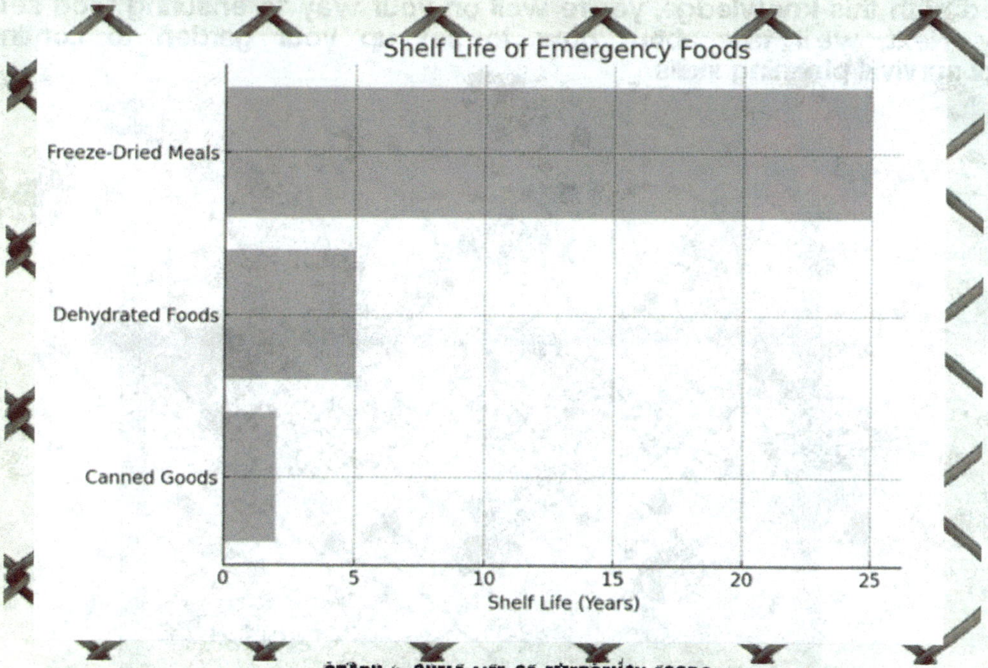

GRAPH 1: SHELF LIFE OF EMERGENCY FOODS
A BREAKDOWN OF THE SHELF LIFE FOR VARIOUS TYPES OF EMERGENCY FOODS INCLUDING CANNED, DEHYDRATED FOODS, AND FREEZE-DRIED MEALS.
BASED ON GENERAL FOOD SAFETY GUIDELINES.

3.8 RESOURCE VAULT

Here is a list of valuable tools and guides to help you improve your food stockpiling skills. These books, websites, and tools offer useful information and practical advice.

Check out Daisy Luther's The Prepper's Pantry: Build a Nutritious Emergency Food Supply, which provides thorough guidance on building and maintaining a stocked pantry.

You can download one of your bonuses at the end of this book now for a handy summary with food preservation tips.

If you are particularly interested in canning, the Ball Complete Book of Home Preserving by Judi Kingry and Lauren Devine is an excellent resource.It has plenty of recipesand step-by-step instructions for canning various foods, so it's great for both newbies and seasoned canners.

The National Center for Home Food Preservation (nchfp.uga.edu) provides current guidelines and step-by-step instructions on preservation methods, including canning, dehydrating, and fermenting.

The Excalibur Dehydrator is considered to be efficient and reliable for dehydrating fruits, vegetables, and meats.

Another great tool is the All-American Pressure Canner, which is known for its capacity and durability, and is a favorite among homesteaders for canning low-acid foods.
To stay organized, consider using the Pantry Check app. It helps you track inventory and expiration dates,ensuring your pantry remains well-stocked and up-to-date.

For budgeting, the YNAB (You Need A Budget)app is excellent for setting and maintaining a budget dedicated to building your pantry.

In summary, we've explored methods to build a well-stocked pantry, preserve food through canning, dehydration, fermentation, and curing, and the benefits of smoking meat. Armed with this knowledge, you're well on your way to ensuring food security for your family. Next, we'll talk about how to set up your garden to continue your foundational survival planning skills.

GARDENING

> The best time to plant a tree was 20 years ago. The second best time is now.
>
> **CHINESE PROVERB**

Step out of your back door, and instead of a patch of grass, you see a garden full of life with rows of vegetables, herbs, and fruit trees. With a bit of planning and effort, you can actually make this happen. Gardening off the grid is a significant step towards sustainable living and brings a great sense of satisfaction and security. When the grocery store shelves are bare, that potato you grew will taste like victory.

4.1 STARTING YOUR OFF-GRID GARDEN

When picking a spot for your garden, make sure it gets 6-8 hours of direct sunlight every day. Take a walk around your property at different times to see how the sun moves, and avoid shaded areas.

Soil quality is another critical factor. Testing the pH levels and nutrient content of your soil will give you a clear picture of its health. A soil pH between 6.0 and 7.0 is ideal for most garden vegetables. You can use a simple soil test kit from a garden center or send samples to a local agricultural extension service for a more detailed analysis.

Testing your soil's fertility and adjusting its composition is a step you shouldn't skip. If testing yourself, collect soil samples from different spots in your garden, avoiding atypical areas, like where pets might have relieved themselves or where water pools after rain. Use a stainless-steel trowel or chrome-plated tool to prevent contamination. Mix the samples in a clean container and follow the instructions in your chosen soil test kit.

The test results will tell you whether your soil is acidic, neutral, or alkaline and what nutrients it might be lacking.

This information helps you amend the soil with the right fertilizers and organic matter to create optimal growing conditions like lime to raise pH or sulfur to lower it. This will ensure your soil provides the correct nutrients your plants need for robust growth.

There is a way to make your own pH test using cabbage leaves. To make your own pH test:

- Boil two to three red cabbage leaves in one to two cups of water for 5 minutes.
- Allow to cool for 15 minutes.
- Create a soil sample by taking a handful of sieved soil and placing itin a container with a cupful of distilled water or natural rainwater until runny.
- Leave for 30-60 minutes.
- Strain using a coffee filter or folded muslin cheesecloth.
- Place samples into several white or clear containers, jars, dishes, or bottles are ideal.
- Keep one container for your cabbage water as a control to compare your sample's color against.
- Add a dessert spoon of the red cabbage water to one of the soil samples, stir it, and let it sit for about 10 minutes. Then, check for any change in color and compare it to your untouched cabbage water. If the sample's color turns pinker than your control, your soil is acidic. If it turns more blue-green than your control, it's alkaline. You may want to test several soil samples by repeating this same process.

Access to water is vital for your garden's success. Having a reliable water source nearby, like a well or stream, can make a big difference. Instead of carrying water around, you can use an irrigation system with poly tubing, drip lines, or soaker hoses. It saves time and effort and makes sure your plants get the right amount of water without any waste.

Collecting rainwater in barrels and using it for your garden will both conserve water and provide a sustainable and cost-effective solution. We'll talk about this in the chapter on water filtration and collection.

Creating a physical barrier and using natural repellents can save your crops from being devoured by critters. Fences can keep larger animals like deer and rabbits out, while netting can protect your plants from birds. Natural deterrents like planting lavender, rosemary, or marigolds around the perimeter can help keep smaller pests at bay. Companion planting, for example, growing basil with tomatoes, can also deter insects.

Gardening off the grid is rewarding, offering both immediate and long-term benefits. By carefully choosing the perfect spot for your garden, testing and improving your soil, making sure you have reliable access to water, and keeping pesky pests away, you're creating a resilient, sustainable food source that can see you through any emergency. It is just so amazing and satisfying to sit down to a dinner that didn't come from the store but from your own backyard! No need to wait in grocery lines either: just pure, homegrown goodness.

> Going back to a simpler life is not a step backwards.
>
> JOSEPH ADDISON

4.2 PREPARING YOUR SOIL

When you're getting ready to plant, it's important to take care of your soil by adding good quality compost and organic fertilizers. Composting your kitchen scraps, yard waste, and other organic materials will provide essential nutrients to improve the overall quality of your soil. And if you need to, you can build your own compost bin using wooden pallets or wire mesh to create a contained area. Then toss in your veggie scraps and dried leaves in layers, and don't forget to give the pile a good turn now and then to help it break down faster.

You can improve soil structure and fertility using several methods.

1. We want to add compost or well-rotted manure to improve the soil texture and increase water retention.
2. Use wood chips or straw to mulch your garden and keep the soil moist while preventing weeds.
3. Try planting cover crops such as clover or rye, which can later be tilled back into the soil to add organic matter and nutrients.
4. Test soil pH and nutrient levels to give you a guide on whether to add sulfur or lime to adjust pH and how to balance fertilizers to correct nutrient deficiencies.

Caring for your soil in these ways will ensure it is in top conditionfor planting, giving your crops the best possible chance to thrive.

4.3 SELECTING THE RIGHT CROPS

IMAGE COURTESY OF BASIER

Choosing suitable crops for your survival garden is key to ensuring high yields, nutrition, and ease of growth.Start with the basics: heirloom versus hybrid seeds. Heirloom seeds have been passed down through generations, keeping their genetic diversity intact. They're open-pollinated, so you can save and replant them year after year, making them a solid, sustainable option for your garden. Hybrid seeds are bred for specific traits like disease resistance or higher yields but are often sterile and can't be saved for future planting. Weighing these options helps you decide what works best for your garden's long-term sustainability.

Root vegetables like potatoes, carrots, and beets are excellent choices for a homesteader garden. They are nutrient-dense, store well, and do very well in different climates. You can easily grow potatoes in containers or right in the ground, depending on what works best for your space. Carrots and beets are tough enough to handle cooler weather so that you can get them in the ground early for a head start on your harvest. Leafy greens such as spinach, kale, and lettuce are quick-growing, packed with vitamins, and can be harvested many times throughout the season, so you get the benefit of a continuous supply of fresh greens.

Legumes like beans and peas are not only nutritious but also beneficial for soil health. They fix nitrogen in the soil, improving fertility for future crops. Beans can be grown on trellises to save space, while peas thrive in cooler temperatures and can be planted early. Crop rotation helps maintain soil health by preventing nutrient depletion and reducing pest build-up. For instance, following legumes with heavy feeders like corn or tomatoes can optimize soil use. reducing pest build-up. For instance, following legumes with heavy feeders like corn or tomatoes can optimize soil use.

CALENDULA WITH TOMATOES TO DETER PESTS
IMAGE COURTESY OF CAROLGOMEZ

As we mentioned, companion planting enhances growth and deters pests. Planting basil with tomatoes can improve flavor and repel insects. Marigolds deter nematodes and attract beneficial insects. The "Three Sisters" method (growing maize, climbing beans, and winter squash together) illustrates effective companion planting. Maize provides support for beans, which in turn fix nitrogen in the soil, and squash shades the ground, reducing weeds. You can download a useful summary table of companion plants and other gardening tips as part of your free bonuses at the end of this book.

Seasonal planting ensuresyear-round harvests. Plan your garden to include cool-season crops like broccoli and cabbage in early spring and fall and warm-season crops like peppers and tomatoes in summer. This approach maximizes your garden's productivity and diversifies your food supply.

4.4 INDOOR GERMINATION VS DIRECT SOWING

Deciding whether to start seeds indoors or sow them directly outside depends on several factors. Indoor germination is ideal for plants that need a longer growing season or a more controlled environment. By starting seeds indoors, you can extend the growing period, ensuring your plants are strong enough to withstand outdoor conditions. This method is particularly useful for crops like tomatoes, peppers, and herbs. On the other hand, direct sowing is suitable for hardier plants that can tolerate cooler temperatures and have faster germination times, such as peas, beans, and root vegetables.

When starting seeds indoors, you'll need seed trays, grow lights, and heat mats. Fill the trays with a seed-starting mix, pop the seeds in at the recommended depth, and keep them consistently moist. Grow lights should be positioned a few inches above the seedlings and adjusted as they grow. Heat mats can help maintain the ideal soil temperature for germination. Once the seedlings have developed a few sets of true leaves, it's time to think about transplanting them outdoors. Timing is important: ensure the risk of frost has passed and the soil temperature is suitable for the specific crop.

Transplanting seedlings requires a process called 'hardening off'. Gradually acclimate the seedlings to outdoor conditions by bringing them outside for a few hours each day, increasing the time over about a week or so. This helps them adjust to temperature fluctuations, wind, and direct sunlight. When transplanting, dig holes that are slightly larger than the root ball, place the seedlings in the ground, and firm the soil around them. Water them well to help reduce transplant shock.

Direct sowing involves planting seeds directly into garden soil. It's essential to follow the recommended depth and spacing for each crop. For instance, carrots should be sown about a quarter inch deep and spaced two to three inches apart. Regular watering is essential, especially during germination. Using raised beds can improve drainage and soil quality, making it easier to control weeds and pests. Building simple irrigation systems, like drip lines or soaker hoses, ensures your garden receives consistent moisture without wasting water.

4.5 MAINTAINING YOUR GARDEN

Maintaining your garden requires a steady routine of organic fertilization, mulching, watering, and weeding.Organic fertilizers like compost and manure provide key nutrients without the harmful chemicals found in some synthetic options. Mulching with straw or wood chips helps retain moisture and suppress weeds, making your gardening efforts more efficient. Regular watering is fundamental, but consider using conservation methods such as collected rainwater or drip irrigation. Weeding is another task that, while labor-intensive, keeps your plants healthy by reducing competition for nutrients and water.

Pest control is a significant aspect of garden maintenance. Synthetic chemicals are not an option for off-grid, sustainable living, so natural methods are your best bet. Beneficial insects like predatory beetles and ladybugs can keep harmful pests in check. Encouraging these insects involves planting flowers like marigolds and daisies, which attract them. You can rely on companion planting as another effective strategy. For example, planting garlic alongside your vegetables can repel aphids and beetles.

Neem oil is an excellent natural pesticide. It disrupts the life cycle of insects, making it difficult for them to grow and reproduce. A simple spray made from neem oil and water can deter common garden pests like aphids, whiteflies, and spider mites. Apply it to affected plants, ensuring you cover both the tops and bottoms of leaves. This natural method keeps your garden pest-free without harming beneficial insects or the environment. Act immediately to stop pest infestations and diseases from spreading.

4.6 RESOURCE VAULT

To help you learn more about off-grid gardening, I wouldl ike to share several invaluable tools with you.

Gaia's Garden: A Guide to Home-Scale Permaculture by Toby Hemenway offers practical insights into sustainable gardening practices with step-by-step guides and real-life examples. For a more technical approach, The Vegetable Gardener's Bible by Edward C. Smith is another excellent resource. It covers everything from crop rotation to pest control with easy-to-follow instructions.

Gardening off the Grid is a fantastic website that covers essential topics like soil quality, water conservation, and pest control. The site also offers interactive tools and community forums where you can have your questions answered and swap experiences with other off-grid gardeners.

Another great online resource is the Old Farmer's Almanac Companion Planting Chart. This chart helps you understand which plants grow well together ,improving your garden's resilience and productivity.

Tools are just as important as knowledge. A high-quality soil test kit like the Luster Leaf 1601 Rapitest helps you monitor soil pH and nutrient levels. For watering, consider using a Dramm 12380 Heavy-Duty Brass Adjustable Hose Nozzle for its durability and efficiency. A sturdy wheelbarrow, such as the Jackson M6T22, makes transporting soil and compost easier.

Now that we have covered some of the garden essentials, let's turn our attention to water purification and storage methods, which we will cover in the next section

Water is the source of life, andthe symbol of purity and clarity.

ANTOINE DE SAINT-EXUPERY

COLLECTING AND FILTERING WATER

Water is indeed life, and in a survival situation, it becomes even more critical. Picture yourself hiking through the wilderness, miles from civilization, and you come across a clear stream. It looks pristine, but drinking unfiltered water can be a dangerous gamble. Contaminants invisible to the naked eye can pose severe health risks. This chapter is dedicated to ensuring you never have to take that gamble. Understanding water filtration isn't just for the wilderness; it's essential for off-grid living, where access to clean water can be a matter of life and death.

5.1 HOW TO MAKE A DIY WATER FILTER

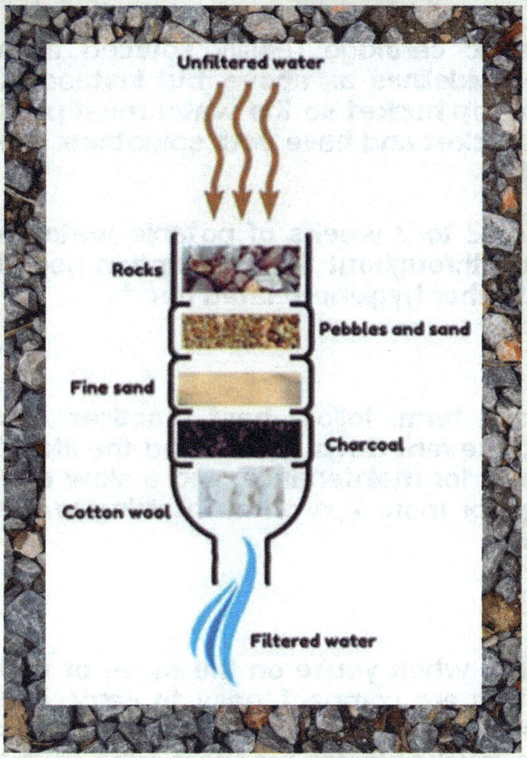

Untreated water can harbor a host of contaminants that pose significant health risks. Nobody wants an encounter with bacteria like E. coli, viruses such as norovirus, or parasites like Giardia. We want to make our water cleaner than our conscience after a good Netflix binge-watch. Physical impurities like dirt and sediment might seem harmless but can lead to digestive issues and inhibit the effectiveness of other filtration methods. Chemical pollutants, including pesticides and heavy metals, can have long-term health effects. Knowing how to filter water effectively is essential for maintaining health and well-being.

Must-do tips for preppers include always carrying a backup water filter and knowing how to improvise filtration methods using available materials. It is relatively easy to create your own filtration system. The idea is to pour your unfiltered water through several layers of readily available materials. Having multiple filtration options is vital so you're not stuck relying on just one system. Each type of filter has its strengths, and understanding how to put them together will massively enhance your water security.

One easy method involves using a 2-liter plastic bottle with the bottom cut off to filter small amounts of water effectively. Start by placing a coffee filter (and something to tie it around the neck) or cotton wool inside the neck of the bottle, and then fill the upside-down bottle with layers of activated charcoal, (clean) sand, gravel, and rocks. The rocks, gravel, and sand are the top layers that trap particles. You can purchase activated charcoal or make it yourself by heating hardwood in a low-oxygen environment until it turns into charcoal. This process creates charcoal with increased surface area, which adsorbs contaminants more effectively. After creating or acquiring the charcoal, crush it into small pieces before adding it to your filter.

This layered approach traps particles, removes chemicals, and improves the taste of the water. A filter layer using broken-up parts of ceramic is another option, using tiny pores to filter out bacteria and protozoa; however, check the source of the ceramics you use, as some ceramics are made using lead, which could be harmful. To make sure your water is safe, it's best practice to pair your DIY filtration with boiling or water purification tablets since the filtration alone might not remove all microorganisms.

Building a larger gravity water filter with easily accessible materials is straightforward and also highly effective, but it follows a slightly different process from the smaller DIY filters. Gather two food-grade buckets and assemble your filter materials from what you have available to you: activated charcoal, fine and coarse sand, gravel, and pebbles. Place one bucket on top of the other, drill a hole in the bottom of the upper bucket for the filter cartridge and another hole near the base of the lower bucket for a spigot. Ideally, you want to pre-filter the water through something to take out the larger particles, such as cheesecloth. Layer the materials in the upper bucket, starting with a layer of material to prevent smaller particles from getting through, then add a layer of (clean) fine sand, followed by coarse sand, then activated charcoal, then gravel, and pebbles at the top to keep everything underneath from getting dislodged. The multiple layers effectively remove particles and contaminants. Let the water run through it and check for clarity and taste before using it. Don't forget to use chemical filtration or boil the filtered water to eliminate any leftover pathogens. Regular maintenance, such as cleaning the filter materials and replacing the activated charcoal, ensures long-term effectiveness.

If you have a store-bought filter, such as a ceramic cartridge (easily sourced from hardware stores or the interweb), follow the same guidelines as above, but instead of the materials, you can set up your ceramic filter in the top bucket so the water must pass through the filter before dripping down to the lower bucket and have your spigotnear the bottom of the lower bucket to let water out. Et voila!

The recommendation is to stock up on a minimum of 2 to 3 weeks of potable water in advance, and then your DIY filter can keep you going throughout the interruption period and provide you with water for drinking, cooking, and other hygiene-related needs.

Troubleshooting Your Water Filter

To keep your filter working efficiently over the long term, follow best practices. For example, pre-filtering particularly dirty water can help prevent clogs and extend the life of your system. If you notice cloudy water, it's likely time for maintenance, and a slow rate suggests layers may be clogged and need cleaning. For more tips on maintaining water filtration, download your bonus at the end of this book

5.2 PORTABLE WATER FILTERS

Access to clean drinking water can be tricky, especially when you're on the move or find yourself in an uncertain situation. Portable water filters are compact, easy to carry, and can ensure you have safe drinking water from natural sources. They don't remove viruses but do remove 99.9% of bacteria. This makes them invaluable for preppers who might need to evacuate or find themselves away from a reliable water source. The convenience of these devices can't be overstated, particularly when you're facing the unpredictable. As a seasoned survivalist, having a portable water filter is a non-negotiable part of your gear.

When it comes to choosing the right portable filter, knowing basic information and costs can help you decide on the right one for you:

Access to clean drinking water can be tricky, especially when you're on the move or find yourself in an uncertain situation. Portable water filters are compact, easy to carry, and can ensure you have safe drinking water from natural sources. They don't remove viruses but do remove 99.9% of bacteria. This makes them invaluable for preppers who might need to evacuate or find themselves away from a reliable water source. The convenience of these devices can't be overstated, particularly when you're facing the unpredictable. As a seasoned survivalist, having a portable water filter is a non-negotiable part of your gear.

When it comes to choosing the right portable filter, knowing basic information and costs can help you decide on the right one for you:

- The LifeStraw, priced between $15-20, filters up to 1,000 gallons and is ideal for individual use. It's simple, easy to use and lightweight. However, its lifespan is limited, there are no storage options and it doesn't filter viruses.
- The SawyerMini, moderately priced at $20-30, offers a high filtration capacity of up to 100,000 gallons and has a number of versatile attachments. Its cons include a slower flow rate and the need for manual operation.
- On the higher end, the Katadyn Hiker Pro costs between $70-90. It's fast, reliable, and excellent for group use, though it's bulkier and requires more maintenance.

Maintenance and lifespan are critical for ensuring your portable filter performs when you need it most. Regular cleaning and back flushing can extend the life of these filters.Each filter's capacity should be monitored, and cartridges or filters should be replaced as needed to avoid any surprises in the field. Additional tips include keeping a small cleaning kit with your filter. Staying on top of maintenance ensures that your filter remains effective, providing you with safe drinking water over the long haul.

The future of water filtration looks promising as advancements in filtration technology improve. Virus filtration, lighter and more compact designs, and increased filtration capacity are on the horizon, eventually making filters more efficient and more user friendly. In time, these new developments will provide us with even more effective and sustainable options.

5.3 COLLECTING AND STORING RAINWATER

Rainwater harvesting offers numerous benefits for off-grid living. It provides a sustainable and environmentally friendly water source, reducing reliance on municipal supplies. This independence ensures you have water during droughts or emergencies and can lead to significant cost savings. Legal considerations are important; check local regulations regarding rainwater collection to avoid fines. Proper system setup prevents contamination, ensuring your water is safe.

Designing an efficient rainwater harvesting system begins with calculating your water needs and collection capacity. Consider household size and daily water usage. Sizing guides are available online to help with this. For a small household, a 55-gallon rain barrel system is a good general guide for use with irrigation and toilet flushing. If you are exclusively relying on rainwater then you need to factor in approximately 100 gallons per person.

Metal or tile roofs and gutters are ideal for collecting rainwater, PVC pipes are ideal to redirect the flow, while rain barrels or storage tanks store the harvested water.

Example of a rainwater collection point, image courtesy of Halfpoint

For installation, start by setting up rain barrels or storage tanks. Connect gutters and downspouts to these tanks, ensuring they are securely attached. Incorporate flush diverters to improve water quality by diverting the initial runoff, which may contain debris, leaves, and contaminants. Having a proper drainage system and overflow management is essential to prevent flooding around your property. This might involve directing overflow water away from foundations or utilizing additional storage options.

Storage solutions vary, including plastic, metal, and concrete tanks. Above-ground tanks are easier to install, while below-ground tanks save space and protect water from temperature fluctuations. Filtration systems, like sediment filters and UV purifiers, ensure good water quality and pumps can distribute the collected water for gardening and household chores. You'll need to regularly clean gutters, check filters, and inspect storage tanks to keep the system free of debris. Seasonal adjustments, like insulating tanks during winter, help maintain functionality.

For those with the budget, commercial rainwater filtration systems offer advanced options. Sediment filters remove large particles, while carbon filters reduce chemicals and odors. UV purifiers kill bacteria and viruses, ensuring safe drinking water. Reverse osmosis systems provide high-level filtration for potable water. Top brands like RainHarvest, 3M Aqua-Pure, and Clean Rain offer complete systems with varying costs. Consider system size based on your water collection capacity and your intended use. Some systems, especially whole-home setups, may need professional installation.

5.4 CREATING A GREYWATER RECYCLING SYSTEM

Recycling grey water is an effective way to make the most of the water you use daily. Grey water, which includes wastewater from sinks, showers, and laundry (but not toilets), can be repurposed for other uses like irrigation. This not only conserves water but also reduces your reliance on municipal supplies, making it a valuable addition to an off-grid lifestyle as part of your game plan to ensure a continuous water supply even during shortages.

Designing a grey water recycling system starts with thorough planning. Identify which sources of grey water you want to capture and where you'll use it. For instance, water from your shower can be redirected to irrigate your garden. Create a simple layout showing how the water will flow from source to destination. Consider the terrain and gravity flow to minimize the need for pumps. A system that uses gravity can be more reliable and easier to maintain, especially in off-grid scenarios.

Installing grey water piping involves setting up a network of PVC pipes and connections to divert water from your household drains to your garden or storage tanks. Begin by installing diverter valves at the drains of sinks, showers, and washing machines. These valves allow you to switch between sending water to the sewer or the grey water system as needed. Ensure all connections are secure and leak-free. Route the pipes to your garden or storage area, using filters to remove large particles and prevent clogging.

It is important to filter and treat grey water for safe and effective use. Simple filtration methods include using mesh screens or gravel filters to remove solids.For added safety, consider biological treatments like constructed wetlands, where plants and microorganisms break down contaminants. Chemical treatments, like adding a small amount of chlorine, can also be used, but should be done cautiously to avoid harming plants. Regularly check and clean filters to maintain your system's efficiency.

When setting up a grey water system, avoid using harsh chemicals in your household products, as these can harm plants and soil. Opt for biodegradable soaps and detergents.

Ensure that grey water doesn't sit stagnant for long periods in order to prevent bacterial growth and odors. Properly label all pipes and components to avoid cross-contamination with potable water systems. Regularly inspect your system for leaks or blockages and address any issues promptly.

By understanding greywater and implementing a well-designed system, you can significantly reduce your water usage and contribute to a more sustainable lifestyle. This system not only conserves water but also provides a reliable source for non-potable uses, which is crucial in off-grid living.

5.5 RESOURCE VAULT

If you are interested in more information on water filtration and collection, I've compiled a list of books, websites, and tools for more knowledge and to offer you a few practical solutions.

The Prepper's Water Survival Guide by Daisy Luther is an excellent read and covers additional details about finding, storing, and purifying water during emergencies.

Information available online is also crucial for staying updated with the latest techniques and technologies. The Wilderness Awareness School's website offers detailed guides on creating survival water filters. The Survivalist Boards (survivalistboards.com) website provides a wide range of discussions and articles on DIY water filtration, rainwater harvesting, and greywater recycling. I highly recommend engaging with the community on their site, where you can learn real-life insights and additional tips from experienced preppers.

For those interested in rainwater harvesting, The Texas Manual on Rainwater Harvesting is an invaluable guide, which provides detailed instructions and illustrations on designing and installing rainwater harvesting systems. It covers everything from basic setups to more advanced configurations, making it suitable for both beginners and seasoned survivalists.

The Katadyn Pocket Water Filter is known for its durability and reliability and comes highly recommended. This filter can process up to 13,000 gallons of water, making it an excellent long-term investment. Another efficient filter is the Sawyer Products SP128 Mini Water Filtration System. This compact, lightweight filter is perfect for on-the-go use and can filter up to 100,000 gallons of water.

Lastly, consider investing in a First Need XLE Elite Water Purifier. This purifier not only filters bacteria, protozoa,and particulate contaminants but also removes viruses and many chemicals. Its versatility and effectiveness make it a valuable addition to any survivalist's water filtration arsenal.

Understanding and implementing these water filtration and collection methods ensures that you and your family remain hydrated and healthy,no matter what challenges arise.

Now that we've covered sustainable off-grid food and water, it's time to explore energy options. Planning ahead for power outages is just ascrucial and will help you maintain a self-sufficient lifestyle.

RENEWABLE ENERGY SYSTEMS

Do you dream of a day when your home is entirely powered by the sun, free from the grid, and immune to outages? This vision isn't just a fantasy for the distant future; it's achievable now. I remember the first time I saw a neighbor install solar panels on their roof.The idea of harnessing sunlight to power everything from lights to appliances was amazing to me. It felt like a new futuristic age had arrived.

6.1 INSTALLING SOLAR PANELS

INSTALLATION OF SOLAR PANELS, IMAGE COURTESY OF ROBERT GOMEZ

Solar power is an essential piece of an off-gridder's kit to remain self-sufficient. Sunlight causes photovoltaic (PV) cells to convert sunlight into electricity by exciting electrons and creating an electric current. This process (the photovoltaic effect) is the magic behind solar power. PV cells are made from semiconductor materials like silicon. As technology advances, newer materials such as perovskite are emerging, showing promise with efficiency improvements from around 3% in 2009 to over 25% today (source: Energy.gov).

Different types of solar panels cater to various needs. Monocrystalline panels are the most efficient, converting 17%-22% of sunlight into electricity. They are made from a single crystal structure, offering high performance but at a higher cost. Polycrystalline panels are less efficient (15%-17%) but cheaper, made from multiple silicon crystals. Thin-film panels, with 10%-13% efficiency, are generally used in commercial settings due to their flexibility and lightweight nature. Understanding these differences helps in choosing the right panel for your needs.

Solar panel efficiency ratings indicate how much sunlight is converted into usable electricity. The most efficient residential panel for 2024 is the Maxeon 7 445-watt panel, with an efficiency of 24.1% (source: CNET). Government incentives can significantly reduce installation costs. In the U.S., the federal solar tax credit allows you to deduct a portion of the installation cost from your federal taxes. Various state and local incentives may also be available, so check with your local government.

Choosing the right solar panels involves understanding local permits and regulations. Ensure you comply with local codes, which may require permits for installation in some states.

Calculating your daily and monthly energy consumption by reviewing your electricity bills will help determine the number of panels needed. Roof-mounted systems are common, utilizing unused space, while ground-mounted systems offer flexibility in placement and angle. Evaluate your roof space and orientation to maximize sun exposure. South-facing roofs are ideal for capturing maximum sunlight.

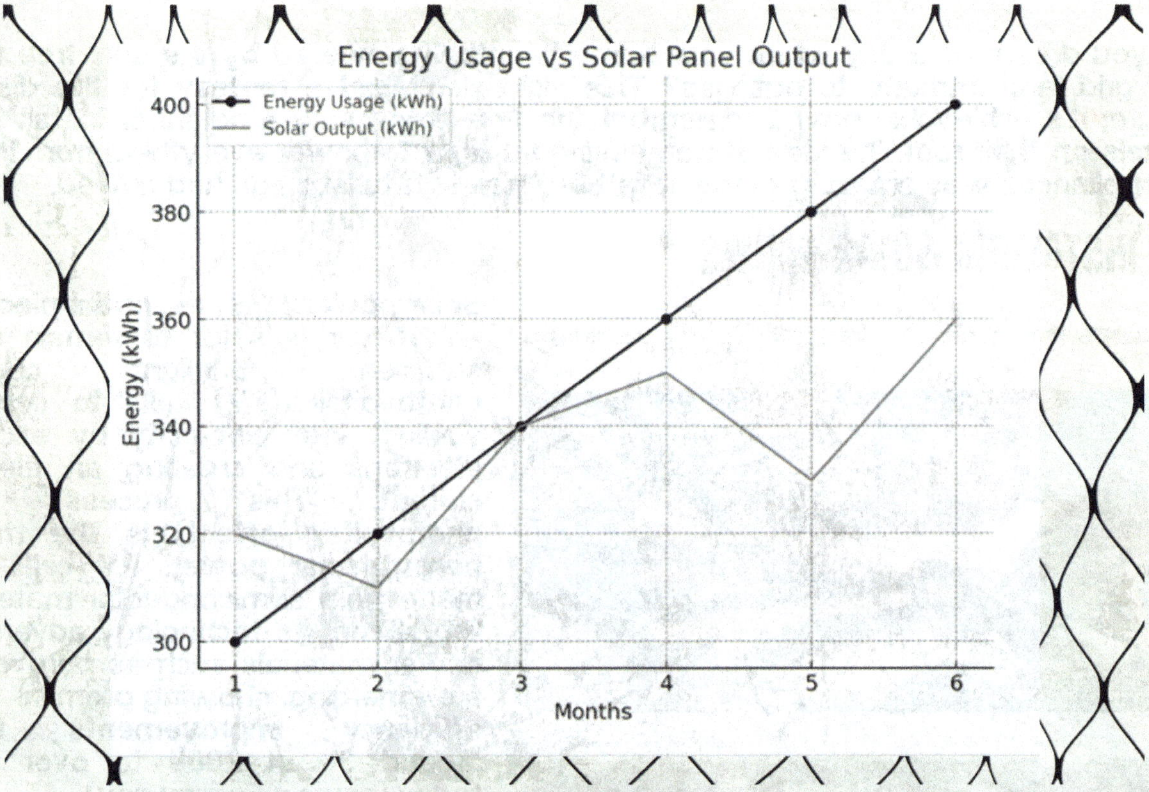

GRAPH 2: ENERGY USAGE VS SOLAR PANEL OUTPUT

A LINE GRAPH COMPARING HOUSEHOLD ENERGY CONSUMPTION TO SOLAR PANEL OUTPUT OVER SIX MONTHS. THIS HELPS ASSESS WHETHER A SOLAR SETUP CAN MEET THE ENERGY DEMANDS OF AN OFF-GRID HOUSEHOLD.
BASED ON GENERAL SUSTAINABILITY GUIDELINES.

When selecting specific brands and models, consider cost, efficiency, durability, and warranty. The Maxeon 7 445-watt panel is currently the most efficient, followed by Canadian Solar's TOPHiKu6 470-watt panel with 23% efficiency. Compare these with polycrystalline panels, which are cheaper but less efficient. For example, installing a 5kW monocrystalline system might cost around $15,000-$25,000, while a similar polycrystalline system could be slightly cheaper. Don't forget to factor in potential savings from government incentives and reduced electricity bills.

For DIY installation, remember that warranties and incentives may not apply. Labor costs amount to only about 10% of the overall cost (source: EnergySage), so professional installation might be worth it. However, if you proceed with DIY installation, start with a site assessment to determine the best location and angle. Gather tools like brackets, inverters, and wiring. Secure brackets and panels to the roof or ground, then connect the panels to the inverter and battery storage systems. Test the system to ensure functionality.

Gather tools ahead of time. Beginners should opt for reputable kits; experts may customize based on specific load needs. You will need:
Solar panels
- Inverter
- Battery storage
- Mounting hardware
- Wiring and connectors
- Drill and drill bits
- Screwdrivers and wrenches
- Ladder and safety harness
- Measuring tape and chalkline
- Multimeter for electrical testing

Roof Mount: Attach panels using mounting brackets or rails, ensuring they're secured at an angle for optimal sunlight.

Ground Mount: Use pole mounts or frames for easy access and to position panels at the best tilt angle.

Connecting the System: Safely wire the panels to the inverter and battery.The process of connecting a battery to an inverter can vary depending on the model and type of battery system you're using. While the general steps (wiring the positive and negative terminals correctly and connecting the inverter and charge controller) will be similar, specific instructions for voltage settings, safety precautions, and wiring configurations can differ based on the manufacturer's design. So always follow the manufacturer's instructions for your particular battery and inverter model to ensure proper installation and to avoid damaging the system.

Finish by testing the system performance, and you'll want to test the performance regularly to keep it operating at full capacity.

Maintenance of solar panels involves regular inspection and cleaning to remove dust and debris. Check for shading issues and panel damage, and inspect connections for loose or corroded wires. Monitoring output with a solar performance system can really help get the best efficiency out of your system.

The field of solar panel technology is rapidly evolving. For instance, companies like Oxford PV are developing perovskite-silicon tandem cells, which combine the strengths of both materials and can achieve efficiencies exceeding 30%, and work is ongoing to circumvent the longevity issues of these panels. Keep an eye on these advancements to stay ahead in the renewable energy game

6.2 WIND TURBINES: SETUP AND MAINTENANCE

Wind energy is a bit of a game-changer for off-grid living.Wind turbines convertthe kinetic energy of wind into electricity, offering a sustainable power source, but you will need to live in the right kind of environment to benefit. The process involves wind turning the blades, which spin a shaft connected to a generator, producing electricity. Mechanically, it's straightforward but requires careful planning and setup. One of the significant benefits of wind energy is reducing your carbon footprint. Wind energy is clean, producing no emissions during operation. Financially, the long-term savings on energy bills make wind turbines a cost-effective investment. You can augment your solar setup with wind power, ensuring you have electricity even on cloudy, windy days. This complementary approach can maximize your energy production year-round.

IMAGE COURTESY OF RANGONI GIANLUCA

When comparing the two systems, solar panels are generally easier to install and yield reliable results, especially in sunny areas. Wind turbines, however, come with the implications of legal planning applications and are best suited for regions with consistent wind. Installing a wind turbine involves navigating local codes and obtaining permits, which can be time-consuming. Yet, wind turbines can produce energy day and night, unlike solar panels.

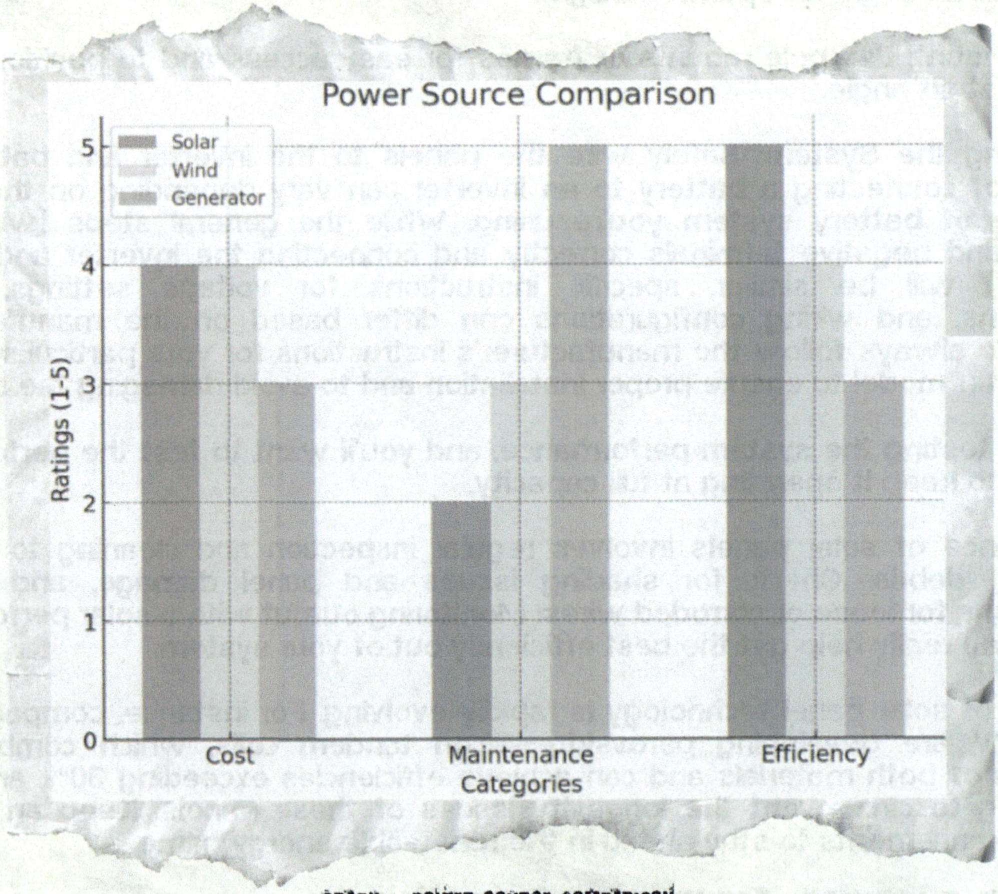

GRAPH 3: POWER SOURCE COMPARISON
A COMPARISON OF SOLAR, WIND, AND GENERATOR SYSTEMS, RATED BY COST, MAINTENANCE, AND EFFICIENCY. BASED ON GENERAL ENERGY PERFORMANCE COMPARISONS.

The energy output and reliability of wind turbines depend heavily on local wind conditions. For places with strong winter winds, wind turbines can be a perfect complement to solar panels, which perform best in summer. It's crucial to assess your circumstances to determine if a wind turbine is advisable.

Choosing the right wind turbine involves several steps. Start by checking online wind maps to calculate the average wind availability in your area. You can also assess this more directly by using an anemometer at the height at which your turbine will be installed.

Wind speeds of 9 to 12 miles per hour are a good range for electricity production. This will give you information on whether it is worth investing in and also in selecting the appropriate turbine. Don't forget you'll also need to find out if you need a permit and how long it takes to obtain one.

There are two main types of turbines: horizontal-axis and vertical-axis turbines. Horizontal-axis turbines are more efficient but require more space. Vertical-axis turbines are easier to install and can handle turbulent wind conditions better. Brands like Bergey and Windstream offer reliable models with varying costs and functionalities. When considering tower options, freestanding towers are robust but expensive, while guyed towers are cost-effective but require more space. Evaluate size and capacity, keeping in mind factors like wind speed, turbine size, cost, and installation requirements.

The installation process for wind turbines is detailed but manageable. Begin with a thorough site assessment to determine the optimal location and height for maximum wind exposure. Assemble the turbine by constructing the tower, nacelle, and rotor blades. Proper wiring is essential; closely follow the manufacturer's instructions and connect the turbine to an inverter and battery storage system, ensuring all connections are secure. Ground the system and secure the tower to prevent damage from high winds. Installing a small wind turbine on a tall tower can significantly increase energy production by capturing higher wind speeds.

Regular maintenance is crucial for long-term performance. Inspect the system regularly, lubricate moving parts, and check for corrosion. You can invest in monitoring tools to track energy output and identify issues like blade damage or mechanical wear. Upgrading components when necessary ensures the system remains efficient. Future trends in wind power will include advancements in turbine efficiency and durability, promising even greater returns on your investment.

6.3 RESOURCE VAULT

I've put together a list of must-have materials if you are exploring sustainable options and to further boost your knowledge and skills in these renewable energy systems.

The Renewable Energy Handbook by William H. Kemp serves as an excellent guide for those looking to integrate renewable energy into their homes. Kemp covers a wide range of topics, including solar, wind, and micro-hydro systems, with clear, step-by-step instructions.

For those specifically interested in solar energy, Solar Power for Beginners: How to Design and Install the Best Solar Power System for Your Home by Lacho Pop MSE simplifies the intricacies of solar panels, making it easy for anyone to understand the basics of design and installation. The book includes detailed diagrams and tips that are particularly useful for DIY enthusiasts.

For online tools, EnergySage (energysage.com) is a fantastic website that offers detailed comparisons of different solar panels, inverters, and installers. The site provides tools to calculate potential savings from solar installations and offers insights into government incentives and financing options.Another great online resource is Windustry (windustry.org), which provides extensive information on wind energy, including feasibility studies, community wind projects, and small wind turbine installation guides.

When it comes to tools, investing in quality equipment will save you a lot of hassle. For solar installations, the Renogy Solar Panel Kit is highly recommended. It includes all necessary components, such as panels, inverters, and mounting hardware, making it a convenient choice for both beginners and seasoned planners. For wind energy, the Pikasola Wind Turbine Generator Kit isa reliable option. It's designed for easy installation and offers efficient power generation, making it suitable for various wind conditions.

Stay on top of monitoring your renewable energy systems to maintain their efficiency. The SolarEdge Inverter, which has monitoring capabilities, is an excellent tool for tracking your solar panel performance. It provides real-time feedback on energy production and system health, allowing you to address issues promptly. For wind turbines, the WindyNation Wind Turbine Controller offers similar monitoring features, ensuring your system runs smoothly and efficiently.

Lastly, don't overlook the value of community and professional advice. Join forums like the Solar Panel Talk or the Small Wind Turbine Forum. These platforms allow you to connect with other preppers and renewable energy enthusiasts, share experiences, and seek advice. Engaging with a community can provide additional support and insights, making your journey toward off-grid living more manageable and enjoyable.

By utilizing these resources, you'll equip yourself with the knowledge and tools needed to implement and maintain renewable energy systems effectively. This not only ensures self-sufficiency but also increases your sustainable and resilient lifestyle.

Let's talk in more detail about battery storage systems to help you maximize the use of your solar and wind power systems.

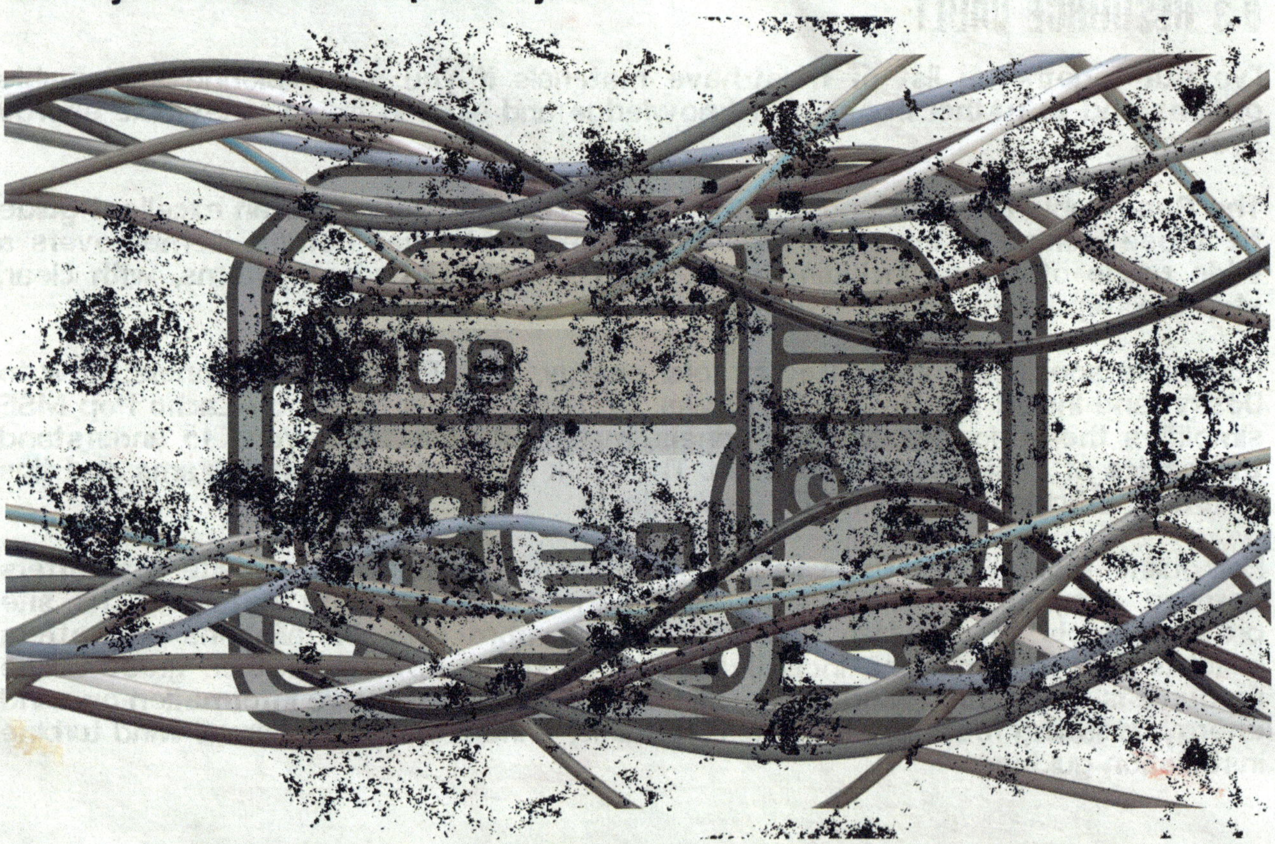

BATTERY STORAGE SYSTEMS

You're listening to the noise of a stormy night when the power suddenly goes out. Your neighbors scramble to find flashlights and candles, but your home remains fully powered. Your lights stay on, your appliances hum along, and your family feels secure. The key to this resilience? A robust battery storage system. These systems are crucial for anyone looking to live sustainably off-grid, as they store and release the excess energy generated by solar panels and wind turbines, giving you a continuous power supply even when the sun sets or the wind dies down.

Electrical system for solar panel installation, battery and inverter
Image courtesy of NOWRAphotography

Battery storage systems harness the excess energy your solar panels or wind turbines produce during peak times and make it available when you need it most.This means you can store solar energy during the day and use it at night or on cloudy days. It's a major advancement for maintaining a consistent power supply and crucial for keeping your household running smoothly in emergencies. So, during a week of overcast weather, your battery storage system can keep your lights on and your refrigerator running, providing peace of mind and security.

When it comes to storing energy from solar panels or wind turbines, choosing the right battery is crucial.

Then you've got NickelCadmium batteries: they last longer and are reliable, but they're pricier and contain some toxic materials, so are less common in residential setups. So keep that in mind as they tend to be more suited to commercial setups. If you're looking for the top-of-the-line option, Lithium-ion batteries, especially the Lithium Iron Phosphate (LiFePO4) type, are your go-to. They're completely maintenance-free, have an impressive lifespan, and are super efficient. They may cost more upfront, but their long life and high performance definitely make them a wiser investment in the long run. They are hybrid systems to accommodate both wind and solar, too.

Sizing and selecting the right battery storage system for your needs involves a few key steps. As with everything, start by working out your energy storage requirements based on your household consumption. Assess your daily energy usage by checking your electricity bills and noting your peak usage times. Compare battery types considering factors like cost, lifespan, maintenance requirements, depth of discharge (DoD), cycle life, and efficiency. For instance, lead-acid batteries should only be discharged to 50% capacity, while lithium batteries can handle 80% or more.

Use an online calculator to estimate your storage needs, factoring in how long the battery will generate electricity for your home. For example, a lithium-ion battery might provide continuous power for 10 hours at full home usage or 20 hours at half usage.

In terms of brands and models, the latest reviews highlight several top choices. The Tesla Powerwall 2, with its 13.5kWh capacity offers high efficiency and a long lifespan, making it a popular choice. The BYD Premium LVS & HVM series is reliable and has a modular design, allowing for flexible capacity expansion over time. The Sungrow SBR Battery is noted for its good value and robust performance. Comparing your battery options is the key to making a smart choice. Take a look at the cost per kWh, how much total energy the battery can store, the power it can deliver, how long it'll last (cycle life), and what type of warranty comes with it. By weighing everything up, you'll be able to find the best option for your needs.

Installing a battery storage system involves careful planning and execution. Begin by organizing the installation site, ensuring it's well-ventilated and protected from extreme temperatures. Safety considerations are paramount. Use proper personal protective equipment (PPE) and follow best practices and the manufacturer's recommendations for handling and installing batteries.

Connect the batteries to your renewable energy system using appropriate wiring, inverters, and charge controllers as directed by the manufacturer. Testing the system is crucial. Check battery connections and electrolyte levels in lead-acid batteries. Use a comprehensive safety checklist to ensure all steps are followed correctly, from securing connections to verifying system performance.

Maintaining and troubleshooting your battery storage system is essential to ensure longevity and optimal performance. Regular inspections can prevent issues like corrosion and loose connections. Monitoring performance through a Battery Management System (BMS) helps track battery health and efficiency. Address common problems such as low capacity, overheating, and voltage fluctuations promptly to keep your system running smoothly so it will last for years to come.

7.1 TROUBLESHOOTING YOUR RENEWABLE ENERGY SYSTEM

Keeping your renewable energy systems in top shape requires regular maintenance. A monthly maintenance checklist can help keep everything in order. This checklist might include checking solar panels for dirt, inspecting wind turbine blades for balance issues, and ensuring battery terminals are clean and connections are secure. Troubleshooting common problems with renewable energy systems is part of the upkeep.

To check how well your solar panels are working, the first thing you should do is give them a quick visual once-over. After that, you can test the voltage output and check for any shading or dirt that might be affecting efficiency. A drop in efficiency can often be traced to shading, dirt buildup, or connection issues, so make sure the panels are clear of any obstructions and give them a good cleaning every few months. Dust, bird droppings, or debris can block sunlight and seriously cut into their output. A simple clean with a soft brush or sponge and water will do the trick. Just be careful not to use anything abrasive that could scratch the surface. Also, take a moment to check all electrical connections, ensuring they're tight and free from corrosion.

Wind turbines require a bit of a different routine. Every six months or so, it's a good idea to inspect the blades, tower, and wiring. Keep an eye out for any signs of wear, rust, or loose connections. Blades can sometimes become unbalanced, which impacts performance. So make sure they're straight and evenly weighted. Noise from the turbine can also signal loose parts or mechanical wear that need attention.

Keeping up with battery maintenance is just as important as choosing the right system. Make it a habit to regularly check charge levels and look for any signs of corrosion on the terminals. If you start noticing problems like your battery holding less charge, getting too hot, or losing voltage, your Battery Management System (BMS) can help figure out what's going on. Overheating might be a sign that your system needs better airflow, while a drop in voltage could mean something more serious is happening with the battery cells.

As your energy needs grow, upgrading and expanding your renewable energy systems becomes necessary. Adding more solar panels or wind turbines can increase your energy production. Upgrading your inverters and charge controllers is a great way to boost efficiency, especially if you're managing a larger system. Upgrading to newer technologies like smart grid systems can really enhance your energy management. These systems provide real-time insights, allowing you to see how much energy you're using and giving you more control over it. By using advanced sensors and software, smart grids can anticipate your energy needs and adjust the power supply accordingly, helping you use energy more efficiently. Looking ahead, we'll likely see more AI-powered energy management systems and even better options for storing energy.

Improving overall energy efficiency is another way to maximize the output of your renewable systems. Insulating homes and buildings reduces energy loss, keeping your interiors warmer in winter and cooler in summer. Using energy-efficient appliances also cuts down on consumption. Practical energy-saving practices for homesteads include sealing drafts, using LED lighting, and setting thermostats to eco-friendly temperatures. Regularly maintaining your systems and staying updated with the latest advancements will ensure you make the most out of your renewable energy setup.

7.2 RESOURCE VAULT

To deepen your understanding of battery storage systems and their role in renewable energy setups, it's essential to have access to quality guides.

The Solar Electricity Handbook by Michael Boxwell is a user-friendly guide and provides detailed instructions on setting up solar systems, including battery storage. Boxwell's step-by-step approach makes it accessible even if you're new to the subject.

For those who prefer online tools, the Clean Energy Reviews website (cleanenergyreviews.info) is invaluable. It offers up-to-date reviews on the best home solar battery systems, including detailed comparisons of brands like Tesla Powerwall 2, BYD, and Sungrow. The site breaks down each system's capacity, efficiency, and cost, helping you make informed decisions. This resource is particularly useful for understanding the trade-offs between different battery types and models.

Another excellent online resource is Unbound Solar (unboundsolar.com). This site provides in-depth articles comparing lead-acid and lithium batteries, discussing their pros and cons in detail. The information here can guide you through the decision-making process, helping you choose the right battery for your needs.

When it comes to tools, it's worth investing in a quality Battery Management System (BMS) like the Victron Energy BMV-712 Smart Battery Monitor. This handy device lets you keep an eye on your battery's performance, helping it run more efficiently and last longer. It gives you real-time information on things like voltage,current, and charge levels, which can help you spot potential issues early before they turn into bigger problems.

For maintenance and troubleshooting, Battery University (batteryuniversity.com) is an excellent resource. The site is packed with useful guides on everything from basic battery care to solving common issues. Whether you're dealing with low capacity, overheating, or voltage drops, you'll find clear, practical advice to help extend your battery's life.

Next, we'll briefly cover non-renewable energy options for during outages, because you never know when systems will fail and you need a backup option.

NON-RENEWABLE ENERGY

The grid goes down and your solar power is out of action, your battery is depleted, and there is no wind, leaving you in the dark with no way to keep your nice leftovers from yesterday from spoiling or to charge your devices. Having a backup power source isn't just a luxury; it's a necessity. When the grid fails, for whatever reason, we need multiple alternatives to keep our homes running. When the lights go out, a generator can keep things moving. It's not fancy, just fuel and a motor, but it'll give you the power you need when nothing else works. It's your safety net.

8.1 GENERATORS

Portable gasoline generator. Image courtesy of ASphotofamily from ASphotostudio

In crises, having a generator can be the difference between comfort and chaos. It can keep the fridge running and the Wi-Fi alive. So, understanding the types of generators and their applications is going to help you if SHTF. You may already know some basics, but let's get into some specifics.

Let's look at a few different types of generators:

Portable generators are really versatile and can be moved around as needed. They're great for short-term power needs and smaller loads.

Standby generators, on the other hand, are installed as a permanent fixture and automatically kick in when the power goes out. They're ideal for whole-house power and long-term outages.

Inverter generators are a bit different. They produce cleaner power, making them suitable for sensitive electronics like computers and medical equipment.
When it comes to fuel types, each has its pros and cons.

- Gasoline generators are common and relatively inexpensive, but gasoline is highly flammable and has a short shelf life.
- Diesel generators tend to come with a higher price tag, but they often provide better fuel efficiency and longer engine life, which can make them a solid investment in the long run. Plus, with diesel fuel being less flammable than gasoline, storing it is generally safer.

- Propane generators burn cleaner than both gasoline and diesel, emitting fewer greenhouse gases. They require an external tank, which can be an advantage or a drawback, depending on your setup.

First things first, you'll want to size your generator properly. Start by figuring out how much power your home really needs. It's simple to do by making a list of your essential appliances along with their wattages and then adding everything up.

Don't forget about those startup power surges. Some appliances, like air conditioners and refrigerators, need more juice to kick on than they do to keep running. A good tip is to choose a generator that can handle a bit more than your total wattage. So, for example, if your total comes to 5,000 watts, it's smart to go for a generator that can produce at least 6,000 watts. This gives you a nice buffer for those surges and helps account for any efficiency losses down the road.

Budgeting for a generator involves several factors. Portable generators can range anywhere from around $500 to $2,000, depending on their power output and features. Standby generators are more expensive, typically costing between $5,000 and $15,000, including installation. Inverter generators fall somewhere in between. Consider your specific needs, how often you expect to use the generator and your budget. While not eco-friendly, investing in a reliable generator can save you from costly disruptions during emergencies.

Proper placement of your generator is vital for safety and efficiency. They should be placed outside, away from doors, windows, and vents, to prevent carbon monoxide poisoning. They need to be on a dry, level surface and protected from the elements. For standby generators, a concrete slab is often required. Portable generators can be placed on a sturdy platform. Ensure the location is easily accessible for re-fueling and maintenance. Avoid placing generators in garages or enclosed spaces, even with the doors open.

When it comes to storing fuel for your generator, a bit of planning is important. Always use approved safety cans for gasoline and keep them outside your living areas for safety. Diesel fuel is generally safer to store since it's less flammable, but it should still be kept in proper containers.

For propane generators, aim to have enough fuel to run for at least a week during an outage. Most propane generators use about 0.5 to 1.5 gallons of propane per hour, so for a week, having around 50 to 100 gallons stored would be a good idea. Just make sure to keep those propane tanks upright in a well-ventilated area. And don't forget to rotate your fuel supply regularly to ensure it stays fresh for when you need it.

Safety is paramount when operating generators. Follow the manufacturer's instructions and never operate a generator with wet hands or in damp conditions to avoid electrocution. Turn off and cool the generator before refueling to prevent fires.Use heavy duty, strong, outdoor-rated extension cords to connect appliances directly to the generator. Avoid backfeeding by not plugging the generator into a wall outlet, as this can cause serious hazards. Install carbon monoxide alarms in central locations and outside sleeping areas to detect dangerous gas levels.

To maintain your generator and improve its longevity, regularly check oil levels, replace air filters, and inspect spark plugs. Run your generator periodically, even when not needed, to keep it in good working condition.

For diesel generators, ensure the fuel system is clean and free from water contamination. Proper care and maintenance will extend the lifespan of your generator and ensure it's ready when you need it.

When comparing generators to solar and battery backups, each has its merits. Generators can run continuously as long as there's fuel, making them ideal for long-duration outages. Solar and battery systems, like the Tesla Powerwall 2, are quieter, fume free, and environmentally friendly, but they are limited by storage capacity. Generators generally have a lower cost upfront but higher ongoing fuel costs, while battery systems are more expensive initially but cheaper to run long-term. You need to make your choice depending on your specific needs, budget, and environmental considerations.

8.2 RESOURCE VAULT

When it comes to non-renewable energy solutions, having the right guidance can make all the difference. Here are some books, websites, and tools to help you explore this topic more.

The Prepper's Guide to Off the Grid Power: The Beginner's Guide to Solar, Wind, and Generator Systems for Remote and Emergency Use by Scott Hunt breaks down complex topics into easy-to-understand language. He covers everything from generator types to fuel storage, making it accessible for both beginners and seasoned survivalists.

Another excellent resource is The Homeowner's Guide to Renewable Energy by Dan Chiras. While it focuses on renewable energy, Chiras provides a balanced view that includes the role of generators as backup systems.He explains how to integrate different power sources effectively, ensuring that you have a reliable energy mix.

For online tools, the American Red Cross website offers comprehensive guidelines on generator safety and usage. Their advice, developed with input from the CDC, NFPA, and the U.S. Consumer Product Safety Commission, is reliable and easy to follow. They cover everything from choosing the right generator size to preventing carbon monoxide poisoning.

Family Handyman is another valuable online resource, especially their article on Diesel vs. Propane Generators. They provide an in-depth comparison of these two fuel types, breaking down their pros and cons. This article helps you make an informed decision based on factors like fuel efficiency, maintenance, and environmental impact.

When it comes to tools, investing in several reliable carbon monoxide (CO) alarms is crucial. The First Alert CO710 is a top-rated model that provides continuous monitoring and has a built-in display for CO levels. It's battery-operated,ensuring it works even during power outages.

For fuel storage, the No-Spill 1450 5-Gallon Poly Gas Can is highly recommended. Its design minimizes spillage and makes pouring easy, reducing the risk of accidents.
A sturdy generator cover like the ClassicAccessories StormPro Generator Cover is another valuable tool. This cover protects your generator from dust, moisture, and UV damage when not in use. Keeping your generator in good condition extends its lifespan.

Finally, consider investing in a fuel stabilizer like STA-BIL 22214. This product keeps fuel fresh for up to 24 months, preventing it from degrading and causing engine problems. It's a small investment, but it will save you a big headache in the long run. These resources and tools offer practical advice, safety tips, and maintenance guidelines that are essential for any off-gridder. You'll be well-equipped to handle power outages and ensure your family's safety and comfort.

As we wrap up this chapter, one thing's for sure: having a reliable backup power source is like having a trusty umbrella in a rainstorm: essential. Generators are your go-to allies when the lights go out and the world goes a little sideways. But being prepared is about more than just keeping the fridge running and the Wi-Fi alive. Now, let's shift gears and discuss something just as important: home defense and self-defence strategies. It's crucial to ensure your family's safety, especially in uncertain times. When things get tough, staying safe becomes really important. So next, we'll look at some simple steps youcan take to help you get set and protect yourself and your family.

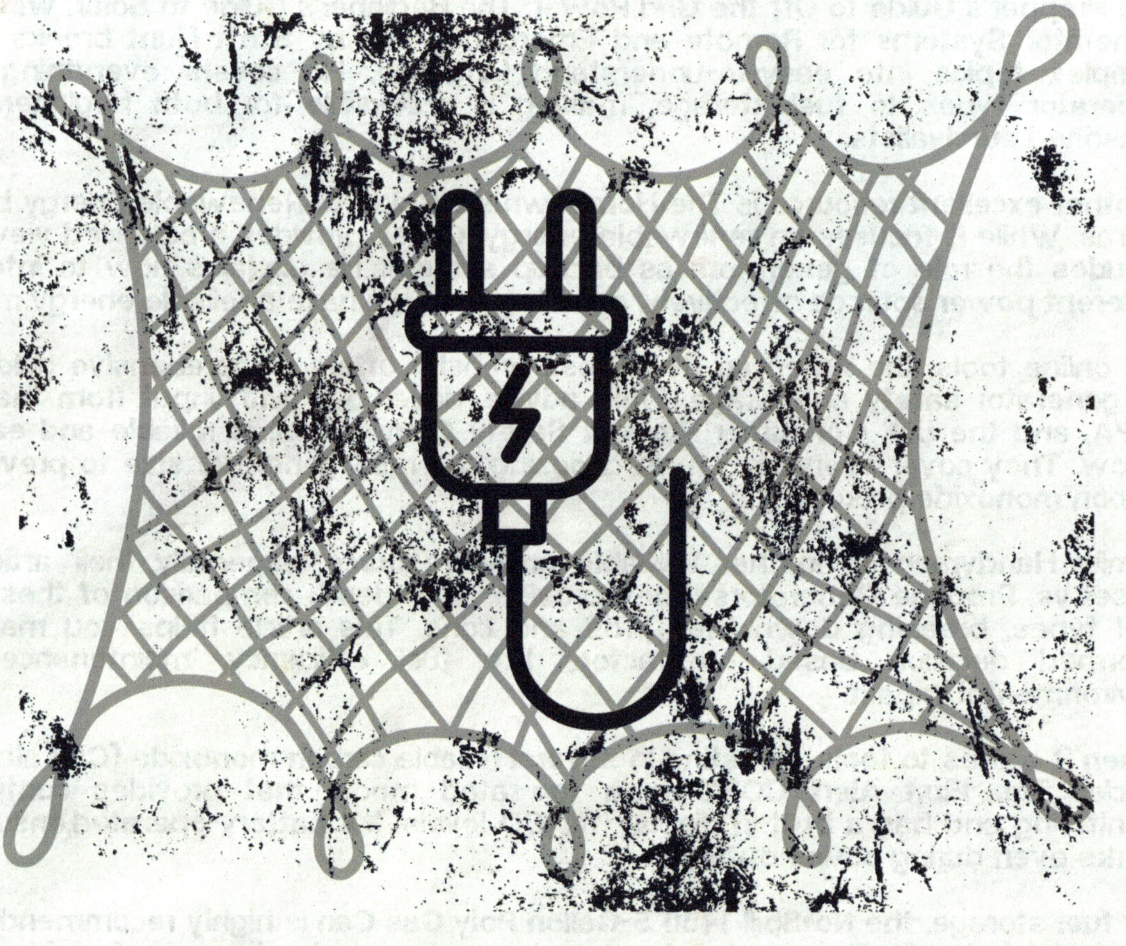

HOME DEFENSE AND SELF-DEFENSE

Water is the source of
life, andthe symbol of
purity and clarity.
Antoine de Saint-Exupery

ABRAHAM LINCOLN

Picture this: it's late at night, and everyone in your house is peacefully asleep. Suddenly, you hear a strange noise coming from downstairs. You quietly tiptoe to checkit out, your heart poundingin your chest, only to find an intruder trying to breakin. This scenario is a homeowner's worst nightmare and highlights just how important it is to have a solid home security system in place. In these unpredictable times, a comprehensive security plan isn't a luxury; it's a must-have, because the only drama we need is in our Netflix queue, not in our homes.

9.1 CREATING A DIY SECURITY SYSTEM

A robust home security system is not just about deterring intruders but ensuring the safety and peace of mind of your loved ones. A layered security approach is crucial.This means not relying on a single line of defense but incorporating multiple security measures to create a fortified home. Think of it as building layers of protection that an intruder would have to breach one by one, making it increasingly difficult for them to succeed.

DIY security systems offer significant advantages over commercial options.For one, they are straightforward to install, allowing you to tailor the system to fit your specific home layout and security needs. Unlike commercial systems, which can be rigid and expensive, DIY systems are highly customizable and scalable. This flexibility means you can start small and expand as your needs change.More importantly, these systems empower you to understand and control your security measures, fostering a sense of self-reliance.

When constructing your security system, several essential components form the backbone.

- Security cameras and surveillance systems are vital.Brands like SimpliSafe and ADT Self Setup are top choices. SimpliSafe is known for its easy installation and compatibility with smart home devices like Alexa and Google Home, with equipment costs starting at $280. ADT Self Setup combines ADT's reliability with Google Nest's high-end security cameras, featuring person and package detection.

- Placement matters:position cameras at entry points, such as doors and windows, and high-traffic areas inside your home.
- Consider placing outdoor camera to cover the perimeter.
- Installation is straightforward: most systems come with mounting hardware and detailed instructions.
- Motion detectors and alarms are another critical layer. They serve as the first alert system, notifying you of any unusual activity. Indoor motion detectors can be placed in hallways and living rooms, while outdoor detectors are best positioned near entry points and along fences.
- Brands like Cove offer affordable and customizable systems, with professional monitoring starting at $17.99 per month. For alarms, consider options like abode, which supports integration with various smart home systems, ensuring comprehensive coverage.
- Reinforced doors and windows significantly enhance your home's security. Brands like Andersen and Pella offer reinforced doors and windows designed to withstand forced entry.
- Installing door and window sensors adds another layer of security, immediately alerting you to any unauthorized entry. These sensors are easy to install and can be connected to your alarm system for real-time alerts.

For a basic alarm system, start with motion sensors in key areas and gradually add more components as needed.

Integrating smart home technology will further enhance your security system. Connecting your security components to your smartphone or control panel allows you to monitor and control your system remotely. This integration provides real-time alerts and the ability to respond quickly to any security breaches.

If you prefer not to be on a crackable system, advanced DIY security measures include installing perimeter defenses like motion-activated lights to illuminate the area when movement is detected, fences,gates, and tripwire systems. Fences and gates create physical barriers, while tripwire systems provide early warning of any unauthorized access. Hidden surveillance spots, like cameras concealed in potted plants or birdhouses, offer discreet monitoring.

Creating a tripwire alarm system is pretty simple and can be a handy way to get an early warning. With just a wireless doorbell acting as the transmitter, you can set up a tripwire to set off the alarm when it's triggered. It's a budget-friendly option that's easy to install and gives you some extra peace of mind about your property.

To keep your security system running smoothly, it's important to stay on top of maintenance and troubleshooting. Routine check-ins ensure everything's working as it should, while software updates keep your system up to date. Replace batteries regularly to maintain a power supply for wireless components. Inspect hardware for wear and tear and promptly address false alarms, connectivity issues, and sensor malfunctions to keep your system functioning reliably.

9.2 TACTICAL NONLETHAL SELF-DEFENSE

Anyone who knows anythingabout protecting their family understands the importance of self-defense training for personal protection. Equipping yourself to defend yourself and your loved ones isn't just about physical readiness; it's about mental resilience, too. The confidence and peace of mind gained through self-defense training are invaluable.

You walk with a certain assurance, knowing you have the skills to protect your family if the situation arises. Avoidance and de-escalation strategies are always your first line of defense. Spotting trouble before it starts can keep you out of harm's way. Staying aware of what's going on around you is crucial. Keep an eye out to stay aware of your surroundings and any potential threats. Real-life situations often highlight just how important this is. Take, for example, avoiding a mugging simply because you notice suspicious behavior and cross the street.

Knowing basic self-defense techniques can make all the difference in a dangerous situation.

- Striking techniques like punches, kicks, and knee strikes are foundational. A wellplaced punch or kick can incapacitate an attacker long enough for you to escape.
- Defensive maneuvers such as blocks, parries, and evasive movements are equally important. They help you avoid or deflect attacks, reducing the risk of injury.
- Less is more: sometimes it's better to give someone a little tap on the nose than go full Rambo. For instance, a simple palm strike to the nose can be incredibly effective in self-defense. It's straightforward and doesn't require much strength, yet it can disorient an attacker, giving you and your family a chance to get away. There are a number of points on the body that can disable an attacker, giving you time to escape, including eyes, ears, nose, throat, groin, knees and solar plexus just beneath the ribs near the diaphragm. You are aiming to disable them just long enough so you can escape.Some of these soft areas can be irreparably damaged so use caution while keeping you and your family safe.

Image courtesy of Pedjasayeret1

Advanced tactical training takes self-defense to the next level. Grappling and ground fighting techniques teach you how to escape holds and submissions, which are crucial if an attacker takes you to the ground. Using improvised weapons can also be a lifesaver. Everyday objects like keys, pens,or even a rolled-up magazine can serve as defensive tools. Techniques for disarming an attacker with a common object can be surprisingly effective. For example, using a sturdy flashlight to strike or block can help fend off an attacker while keeping a safe distance.

Finding and selecting the right self-defense training program is essential for honing these skills. Research local martial arts schools and self-defense courses to find one that fits your needs. Evaluate the instructors' credentials and teaching styles to ensure they are experienced and reputable. Joining a Krav Maga class, for instance, https://www.kravmaga.com/ offers comprehensive tacticalself-defense training. Krav Maga is based on a practical approach to real-world situations, making it an excellent choice for both beginners and advanced practitioners.

Here are some links to get you startedon self-defense:

Here is a link and QR code to a introduction video from https://www.pasquinilli.com/

with some basic self-defense that can work for women - or anyone:

https://www.youtube.com/watch?v=OTAMutleHxQ&t=183s

Here is a handy link from https://www.youtube.com/@BRIGHTSIDEOFFICIAL and QR code to learn about self-defense strategies:

https://youtu.be/T7aNSRoDCmg?si=jxtNd-M52JaG1QtE

SCAN ME

9.3 RESOURCE VAULT

When it comes to fortifying your home and honing your self-defense skills, additional guidance will really help you. Here are some top recommendations for books, websites, and tools that can significantly level up your survivalist skills.

For a comprehensive guide on DIY home security systems,The Complete Guide to Home Security: How to Protect Your Home and Family from Burglars, Thieves, and Other Intruders by David L. Baker is invaluable. It covers everything from basic setups to advanced security measures, offering step-by-step instructions that are easy to follow.

Online platforms like Security.org are also indispensable. Their article Best DIY Home Security Systems of 2024 provides in-depth reviews and comparisons of top brands like SimpliSafe and ADT Self Setup. The site breaks down the costs, features, and ease of installation for each system, helping you make an informed decision. Regularly visiting these kinds of websites keeps you updated on the newest features in home security technology.

For self-defense training, Krav Maga: An Essential Guide to the Renowned Method by David Kahn, a prominent Krav Maga instructor, offers detailed techniques and strategies that are both practical and effective. The book is geared towardsreal-world scenarios, making it a valuable resource for anyone looking to enhance their self-defense skills.

Websites like Krav Maga Worldwide provide a wealth of information, including blogs, training tips, and videos. They also offer a directory of certified training facilities and instructors so that you can find reputable training programs near you.

In terms of tools, the Arlo Pro 4 Spotlight Camera comes highly recommended for surveillance. With its wire-free setup, color night vision, and two-way audio, it's a great choice for user-friendly home security.

For motion detectors, the Ring Alarm Motion Detector really stands out. It's compatible with Amazon Alexa, offers customizable motion settings, and has a sleek design. It is effective in detecting movement and seamlessly integrates with other Ring products.

When it comes to personal self-defense tools, the laser Pulse+ offers a reliable way to incapacitate an attacker from a distance by disrupting neuromuscular control and temporarily immobilizing them. Legal in many places, it's a favored choice for non-lethal defense.

If you are looking for an easy option, the SABRE Red Pepper Spray is known for its range and potency. It's simple to use and highly effective, making it a staple in any self-defense kit.

Combining robust home security with effective self-defense skills will ensure you and your family are well-protected. Moving forward, we'll explore safe spaces, focusing on constructing and maintaining areas within your home that offer ultimate protection.

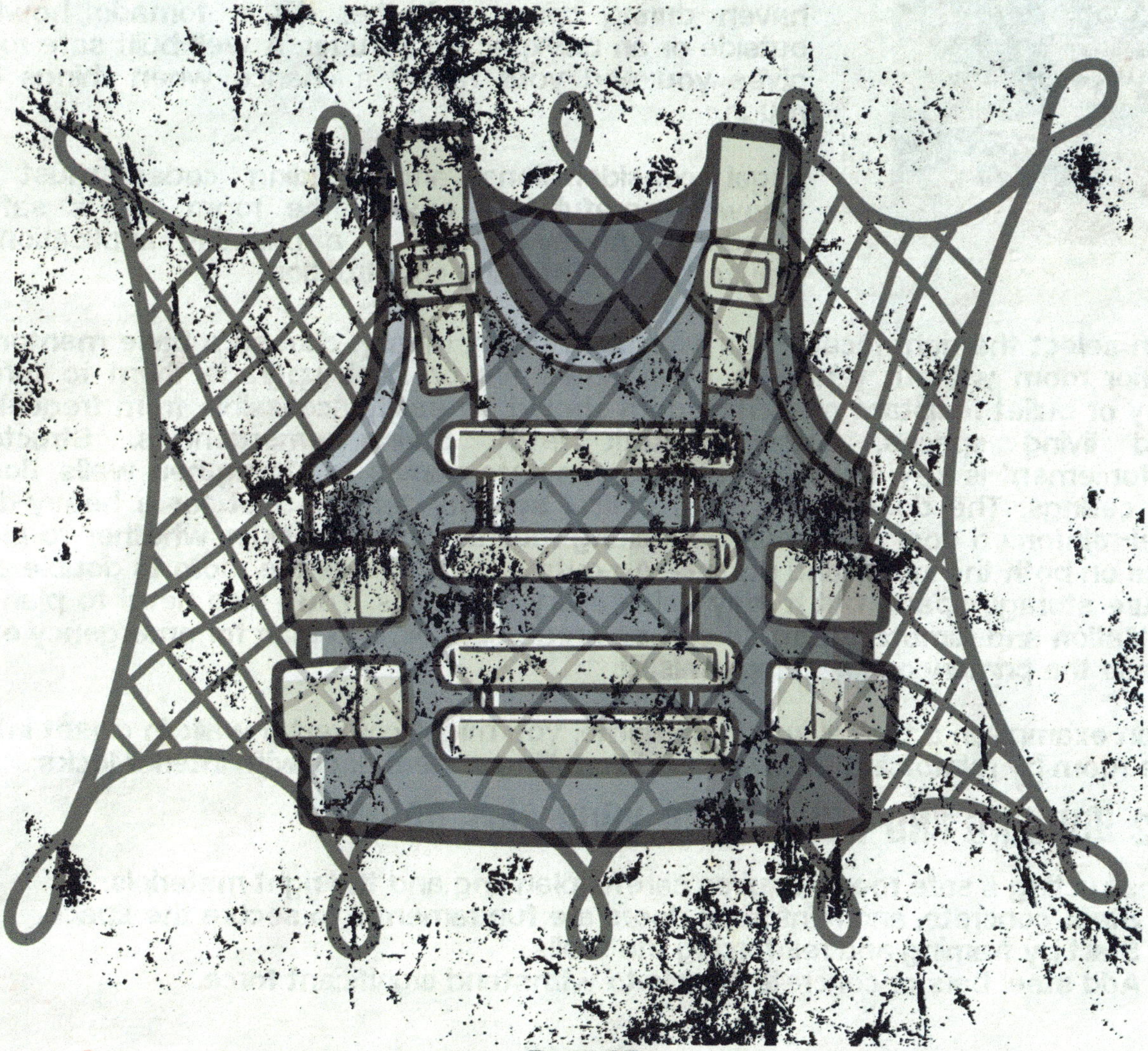

SAFE SPACES

A tornado warning alerts you via your phone. The sky darkens, the wind picks up, and you only have minutes to act.When trouble strikes, having a safe room can be like having a lifeboat in a stormy sea. It's your secure haven during crises, whether it's a tornado howling outside or an unexpected intruder. A well-built safe room gives you and your family a chance when things get rough.

10.1 DESIGNING YOUR SAFE ROOM

A tornado warning alerts you via your phone. The sky darkens, the wind picks up, and you only have minutes to act.When trouble strikes, having a safe room can be like having a lifeboat in a stormy sea. It's your secure haven during crises, whether it's a tornado howling outside or an unexpected intruder. A well-built safe room gives you and your family a chance when things get rough.

Legal considerations and building codes must be followed carefully to ensure the room meets safety standards. Always check if a planning application is required before starting construction.

Then select the right location for your safe room. Ideally, choose a base mentor an interior room without windows. If windows are present, upgrade them to forced entry or bullet resistant ones.The room should be easily accessible from frequently used living spaces, ensuring quick entry during emergencies. Structural reinforcement is key: Use steel bars or concrete panels to strengthen walls, floors, and ceilings. The door is the most critical security feature. Opt for a heavy-duty steel-reinforced door with multiple locking mechanisms. Consider whether to place locks on both the inside and outside. An outside lock allows the room to double as a secure storage space, but it may delay entry in a crisis. You'll also need to plan for ventilation and climate control to ensure breathable air, and plan for emergency exits in case the primary one is compromised.

As an example,in a typical suburban home, you might convert a walk-in closet into a safe room by reinforcing the walls and installing a steel door with interior locks.

10.2 BUILDING AND REINFORCING YOUR SAFE ROOM

Constructing a safe room requires careful planning and the right materials.

- Steel, concrete, and reinforced doors are fundamental to secure the space.
- Start by framing and reinforcing the walls.
- Add steel bars or concrete panels to withstand significant force.

- Installing a secure door is crucial. Choose a heavy-duty steel-reinforced door with multiple locks.
- Ensure the door is secured to the wall with a thick-gauge steel frame.
- Arrange the interior to maximize space and functionality.
- Include shelving for storage and seating for comfort.
- Plan for secondary exits, such as an escape hatch, in case the primary exit is blocked.
- Conduct drills to ensure all family members know how to use the safe room.

Image courtesy of Neshom

The cost of converting a room into a basic safe room can vary: budget for construction materials, labor, and additional features like ventilation and communication systems.

10.3 STOCKING YOUR SAFE ROOM

Properly stocking your safe room is essential for readiness and longterm safety.

- Store non-perishable food and water supplies, aiming for at least three gallons per person and per pet, covering you for up to 3 days.
- Include communication devices like hand crank radios and cell phones with extra batteries and power banks.
- Medical supplies and first aid kits are a must.
- Keep clothes and blankets on hand in case you need to stay put overnight.
- Flashlights and batteries are crucial for power failures.
- Create an inventory checklist to ensure you have all necessary items and regularly update it.

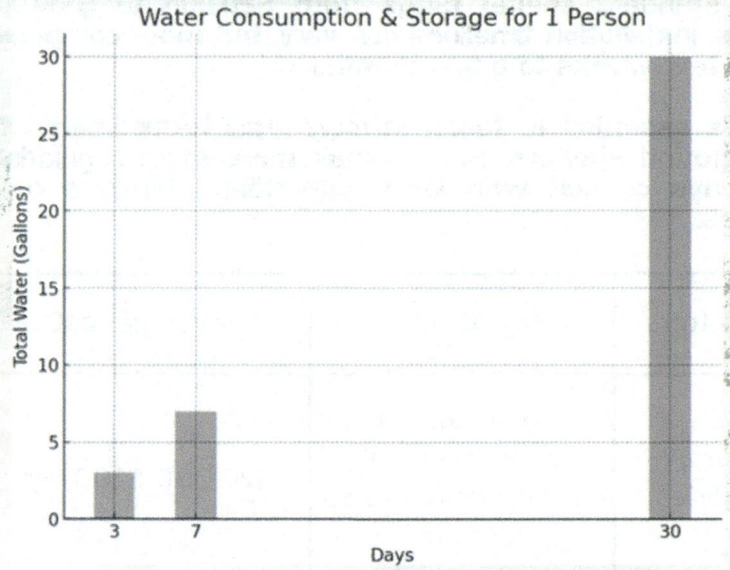

GRAPH 4: WATER CONSUMPTION & STORAGE REQUIREMENTS

THIS CHART SHOWS THE AMOUNT OF WATER NEEDED PER PERSON FOR DIFFERENT DURATIONS (3, 7, AND 30 DAYS), PROVIDING A VISUAL GUIDE FOR HOW MUCH WATER SHOULD BE STOCKPILED.

BASED ON GENERAL WATER CONSUMPTION STATISTICS.

10.4 MAINTENANCE AND UPKEEP

Maintaining your safe room ensures it remains functional and secure. Set reminders for regular inspections to check structural integrity and security features. Rotate food and water stocks every six months and update medical supplies as needed. Conduct family drills to practice using the safe room and review emergency plans regularly. Ensure all family members are familiar with safe room protocols and know what to do in various scenarios. Regular practice and upkeep will keep your safe room set for any emergency.

10.5 SHELTERS AND BUNKERS

Building your own shelter gives you a unique sense of independence and self-reliance that's hard to beat.

Self-build shelters, like small log cabins ,can be a rewarding project.You can source logs locally, which keeps costs down and allows for complete customization. However, the downside is that it requires significant time, effort, and some level of carpentry skills. On the other hand, pre-fabricated kits offer amore straightforward solution. They come with all the necessary materials and instructions, which makes the assembly process way more manageable. Companies like Shelter-Kit and Conestoga Log Cabins provide various options. The downside is that these kits can be costly, and you may need professional help to assemble them correctly.

When it comes to location, building a shelter on your home ground has its benefits. It's easier to maintain and monitor, ensuring its upkeep is manageable. However, having a secondary shelter on an alternative site adds a layer of safety.This is especially useful if your main living space becomes unsafe. The downside is that maintaining a remote shelter can be challenging and costly. Privacy and safety are also concerns; a secluded location may offer better protection from prying eyes and potential threats. Always check local planning application laws to ensure your shelter complies with regulations. Durable shelters like log cabins and pre-fabricated kits often require permits, so it's essential to do your homework before starting construction.

For those with the budget, commercial above-ground and underground survival shelters offer unparalleled protection and convenience. Companies like Atlas Survival Shelters, Rising S Bunkers, Hardened Structures, (and Burrowed.co.uk for those in Europe) provide a range of options. Atlas SurvivalShelters, for example, offers everything from safe cellars to luxury bunkers. They use high-quality materials like bullet-resistant steel and provide features such as NBC (nuclear, biological, chemical) air filtration systems, imported from Switzerland. Average costs for a standard shelter range from $20,000 to $100,000, depending on their size and features. Installation timelines can vary, but most companies aim to complete the project within a few weeks to a few months.

When considering these options, it's essential to factor in local legal requirements for installing above-ground and underground shelters. Some areas have strict regulations that you must adhere to, so always consult with local authorities. Here's a quick comparison of the top four companies:

Company	Types of shelters	Key features	Average cost
atlas survival shelters	safe cellars, platinum series	rould culvert bullet-resistant steel, nbc durable air filters	$20,000-$100,000

company	Types of shelters	key features	average cost
rising s bunkers	bunkers, safe rooms	modular designs, custom finishes	$25,000-$120,000
hardened structures	custom bunkers, safe rooms	high-secuirty, reinforced concrete	$30,000-$150,000
burrowed (uk)	underground bunkers, safe rooms	decontamination rooms, air filtration	$15,000-$80,000

TABLE 2: COMPARISON OF THE 4 TOP SHELTER PRODUCERS. INFORMATION GATHERED FROM LISTED COMPANY WEBSITES.

Atlas Survival Shelters stands out due to its high-quality construction and comprehensive safety features. If budget allows, their shelters offer the most robust protection against a broad variety of threats.

10.6 RESOURCE VAULT

Here's another chance to share a list of valuable materials to enhance your readiness. The Secure Home by Joel Skousen shares the specifics of constructing and reinforcing safe rooms and shelters. His insights into structural integrity and security measures are invaluable.

If you're interested in a more technical approach, Shelters, Shacks, and Shanties by D. C. Beard provides a fascinating look at traditional building techniques. Although it's an older book, the principles it covers are timeless. You'll find detailed instructions on building various types of shelters, from simple lean-tos to more complex log cabins, for those looking to build their own shelters from scratch.

Online sources are also crucial. Fortified Estate's website (fortifiedestate.com) offers a wealth of information on retrofitting existing homes to include safe rooms. They provide detailed guides on selecting secure materials, installing components, and ensuring your safe room meets safety standards.

Another valuable online resource is the Federal Emergency Management Agency (FEMA) website (fema.gov). Their document Safe Rooms for Tornadoes and Hurricanes contains guidelines on designing and building safe rooms that can handle extreme weather conditions, which is especially useful if you're living in an area prone to natural disasters.

If you are starting from scratch, the DEWALT 20V MAX Combo Kit offers a popular set of power tools for any DIY construction project. This kit includes a drill, impact driver, circular saw, and more, providing you with everything you need to build or reinforce your safe room. Another essential tool is the Kreg Pocket-Hole Jig, which simplifies creating strong joints in your building projects.

For those interested in seismic retrofitting, the California Residential Mitigation Program website provides detailed information on strengthening your home against earthquakes. The site also offers financial incentives and grants to help offset the costs of retrofitting.

For those looking to protect their homes from wildfires, the Cal Fire website gives detailed guidelines on creating defensible spaces. The site provides step-by-step instructions on clearing vegetation, maintaining space between trees, and other fire prevention measures. These practices can significantly reduce the risk of your home being engulfed by wildfire.

For those looking into commercial shelters, the Atlas Survival Shelters website (atlassurvivalshelters.com) is a treasure trove of information. The detailed descriptions and customer testimonials provide a clear picture of what you can expect. Their shelters are built to withstand various threats, making them a top choice for those with the budget.

By exploring these options, you'll have access to everything you need to construct and maintain safe spaces and shelters. Next, we'll consider urban survival tactics and focus on practical strategies for those living in city environments where space may be limited.

You may be living in an urban environment where every inch counts. If you live in an apartment with no yard and you yearn for the self-sufficiency that comes with growing your own food, urban gardening offers great solutions. It's entirely possible to reclaim a little bit of nature inside your apartment and turn it into a productive, sustainable source of fresh produce. You can reap many benefits, from reducing your carbon footprint to ensuring a steady supply of nutritious vegetables, regardless of any external crises.Let's explore urban gardening: turning concrete jungles into green oases, one pot at a time.

Urban preppers saw this first-hand in 2025 when wildfires in Florida forced rapid evacuations and gridlock, showing how quickly modern cities can grind to a halt when disaster strikes.

11.1 URBAN GARDENING: MAXIMIZING SMALL SPACES

Thankfully, urban gardening has opened up new possibilities for people in cities. Even if space is tight, you can still grow your own food. It's a great way to become more self-sufficient and have fresh produce literally steps from your dinner table. So instead of relying on store-bought vegetables, you could step out onto your balconyto pick fresh tomatoes or herbs for your dinner. This not only ensures you eat more healthily but also reduces your dependence on the unpredictable availability of supply chains and store-bought goods.

IMAGE COURTESY OF MAICA

The environmental benefits are significant. Urban gardening reduces food miles (the distance food travels from farm to table) which cuts down on your carbon footprint. Growing food locally means fewer trips to the grocery store and less reliance on fossil fuels. Moreover, urban gardens use way less water than traditional lawns. Accordingto a study by LawnStarter, urban gardens can save over 32,000 gallons of water annually for a 1,000-square-foot garden, using 66% less water than lawns. This can be achieved through efficient watering practices and the use of rainwater collection systems, which also contribute to water conservation.

Urban gardening doesn't just stop at environmental benefits; it also uses natural pest control methods. Companion planting, for example, involves growing certain plants together to discourage pests and attract beneficial insects. Planting marigolds alongside your vegetables can deter pests like aphids while attracting pollinators like bees, which are essential for a thriving garden. This eliminates the need for chemical pesticides, making your garden more eco-friendly and your produce a lot safer to eat.

Turning your balcony into a vegetable garden is easier than you might think. Who says you need a yard to grow your own food?

A small balcony, some pots, and a little imagination, and you can have fresh tomatoes, herbs, peppers for dinner, and strawberries for dessert: proof that a green thumb doesn't require a lot of space. Hanging baskets and vertical planters can allow you to grow more in a limited area. This not only provides fresh produce but also enhances the aesthetic appeal of your living space, creating a little green oasis in the middle of a big city.

Urban gardening is versatile. It can be adapted to various spaces, from balconies and windowsills to rooftops and community gardens. For instance, window boxes can be used to grow herbs and small vegetables, so that you can make efficient use of vertical space. Raised beds on rooftops can support a variety of plants, so you can turn unused areas into productive gardens. Community gardens offer another avenue where you can share resources and knowledge with neighbors, fostering a sense of community and togetherness.

Urban gardening is more than a hobby. It empowers you to take control of your food supply and offers a sustainable, self-sufficient way to grow reliable fresh produce, reduce environmental impact, and enhance food security. Whether you have a small balcony or access to a rooftop, urban gardening can transform your living space into a thriving garden. The need for self-sufficiency has never been more apparent, and having a personal garden is a fantastic proactive step toward resilience and food confidence.

11.2 CONTAINER GARDENING

Container gardening is a super practical solution for urban dwellers wanting to grow and eat their own food. The trick is to select the right containers for different types of plants. Short and shallow-rooted vegetables like beans, kale, and spinach are excellent choices. You can use a variety of containers, from five-gallon buckets and terra cotta pots to plastic window planters and even repurposed soda bottles.

Proper drainage and good soil quality are both integral to container gardening. Containers need to have holes at the bottom to ensure excess water can escape, preventing water logging and root rot, and the soil should be rich in nutrients to support plant growth. You can use pre-packaged potting soil, which is designed to provide the right balance of moisture retention and drainage. Adding organic material like mulch and compost can further enhance soil quality. Companies like FoxFarm Soil & Fertilizer Company and Dr. Earth offer excellent organic soil and fertilizer options to get you set up. Alternatively, make your own potting soil by digging up and sieving 70% garden soil, adding in 30% organic matter (sieved homemade compost, leaf mold, or well-rotted manure) and 5-10% sharp sand or grit if you need to add more drainage.

Growing tomatoes and herbs in pots on a balcony can be incredibly rewarding. Start by choosing containers that are deep enough for the roots to grow: at least 12 inches for tomatoes. Place a layer of gravel at the bottom for drainage, then fill it with nutrient-rich soil. Plant your tomatoes and herbs, ensuring they receive plenty of sunlight. Most vegetables and herbs need full sun, so put your containers in the sunniest spot available. Regular watering is essential, as containers dry out faster than ground soil. To maximize space, you can plant herbs and vegetables together in the same container. For instance, basil grows well alongside tomatoes, providing mutual benefits.

Pruning is another critical aspect of container gardening. Regularly prune the top leaves of the main stalk to stimulate plant growth and prevent overcrowding. This practice directs the plant to produce fruit rather than excess foliage. Harvesting times will vary based on your climate and the type of plants you grow. Resources like Cornell University's guide can help determine when your crops are ready for harvest.

Container gardening has a learning curve, but it's a skill worth mastering. It offers a flexible, scalable way to grow food, even in the smallest spaces.With the right containers, soil, and care, you can turn any balcony, patio, or windowsill into a thriving garden. The self-sufficiency and satisfaction of growing your own food are invaluable, especially in uncertain times.

11.3 VERTICAL GARDENING

Vertical gardening offers a smart solution for maximizing space in urban settings. This technique allows you to increase the number of plants you grow without needing extra ground space, and you can transform a bare wall into a lush, productive garden. To get started, gather items like trellises, wall-mounted planters, and hanging baskets. Trellises can be made from materials like wood, plastic, or metal, and they support climbing plants such as beans and peas. Wall-mounted planters and hanging baskets are perfect for growing herbs, strawberries, and even small vegetables. These tools help you utilize vertical space efficiently, turning unused areas into green, productive zones.

Choosing the right type of plants is key to a successful vertical garden. Opt for climbing plants like beans, peas and vining crops such as cucumbers and squash. These plants naturally grow upwards, making them ideal for vertical setups. Herbs like mint and basil also thrive in vertical gardens, adding both flavor and aroma to your home. When selecting plants, consider their growth habits and light requirements. Clearly, you'll have to place sun-loving plants in spots that receive plenty of sunlight, while shade-tolerant plants can be positioned in less sunny areas. This will ensure all your plants get the conditions they need to flourish.

Installing a vertical garden on a sunny apartment wall can be a rewarding project. Start by attaching a sturdy trellis or series of wall-mounted planters. Make sure they are properly fastened to support the weight of the plants and soil. Fill the planters with nutrient-rich soil and plant your chosen crops. Watering is essential, as vertical gardens can dry out quickly. Use a watering wand to reach higher spots, and group plants with similar water needs together. This setup not only maximizes space but also creates a visually appealing green wall that can enhance your living area.

You can grab one of your bonuses at the end of this book now for a quick guide on the best plants for urban gardening and how to get started.

11.4 AQUAPONICS AND HYDROPONICS: INNOVATIVE GROWING METHODS

Aquaponics and hydroponics offer groundbreaking ways to grow food without soil, which is ideal for urban environments.

Aquaponics creates a small ecosystem where fish and plants work together. The fish waste provides key nutrients to the plants, while the plants naturally filter and clean the water, making it healthier for the fish. This cycle boosts overall productivity and cuts down on waste. In a basic setup, you'd keep fish like tilapia or goldfish in a suitable tank. Water from the tank, rich in nutrients from fish waste, is pumped to grow beds where plants absorb these nutrients. The filtered water then makes its way back to the fish tank, completing the cycle.

Hydroponics, meanwhile, involves growing plants directly in water that's been enriched with nutrients. This setup not only saves space but also uses far less water than traditional gardening, making it perfect for city living. You'll often see quicker growth and bigger harvests with hydroponics compared to soil-based methods due to the controlled environment.

Essential components for a hydroponic system include reservoirs, pumps, and grow lights. Systems like Deep Water Culture (DWC), Nutrient Film Technique (NFT), and Ebb and Flow offer various configurations. An average hydroponic setup might cost around $300-$500, including a basic reservoir, pump, and grow light.

IMAGES COURTESY OF PP, AND TOMMY/X

Setting up a hydroponic system begins with choosing the right type. For instance, DWC involves suspending plant roots in a nutrient solution aerated by air pumps. This setupis relatively simple and cost-effective. You'll need a reservoir, air pump, air stones, and net pots to hold the plants.Fill the reservoir with water and add a nutrient solution. Place the plants in net pots and ensure their roots are submerged in the solution. Use grow lights to provide the right amount of light, especially if natural sunlight is limited. Leafy greens and herbs are a fantastic choice for hydroponic systems as they are fast growers and don't require much space.

Creating an aquaponic system requires a bit more complexity. Start by selecting suitable fish species like tilapia, goldfish, or trout, which are hardy and adaptable. You'll need a fish tank, water pump, bio-filters, and grow beds. The fish tank should be large enough to support the fish population and allow for water circulation. Install a water pump to flow water away from the fish tank towards the grow beds, where plants will absorb the nutrients. Bio-filters help break down fish waste into forms that plants can use. Setting up a small-scale aquaponics system in a garage or spare room can be a rewarding project, providing both fresh vegetables and protein in a compact space. Check Rob Bob's Aquaponics and Backyard Farm for tips on how to create a simple homemade system. His YouTube video 'Solids Filter For Aquaponic System Fish Tank Canister Filter' will give you really handy tips for a basic system. The channel also covers aquaponics systems for more advanced homesteaders.

Regular monitoring and optimization help to maintain these systems, including frequently testing pH levels, ammonia, nitrates, and nutrient levels.You may need to make periodic adjustments to keep the ecosystem balanced. Pests and diseases are also something to watch out for. You can tackle these using the organic and non-toxic strategies mentioned later. If you're aiming for a steady supply of food, knowing when to harvest and replant is key. Consider adding automated controls to your setup to simplify maintenance and create the best possible growing conditions for both your plants and fish.

11.5 ADVANCED INDOOR GARDENING TECHNIQUES

Managing nutrient levels in hydroponic and aquaponic systems is critical to keep your plants and fish in good health.In hydroponics, the nutrient solution is basically the lifeline for your plants, so keeping your eye on it is really important. You'll want to check and adjust the pH and nutrient levels regularly. Most plants do best with a pH between 5.5 and 6.5. Be sure to use nutrient mixes made for hydroponics to give your plants the right mix of macronutrients and micronutrients. In aquaponics, it's the fish waste that supplies nutrients, but you've got to make sure things like ammonia, nitrite, and nitrate levels are in check to keep both your fish and plants healthy. Regular testing and tweaking are key to keeping everything balanced. You can bring your water samples to a pet store that sells fish, and they will check the ammonia, nitrite, and nitrate levels for free, or you can buy a kit and easily test the levels yourself.

Pest and disease control in indoor gardening setups requires a proactive approach. Since you're growing indoors, pests are less likely, but not impossible. Use organic and non-toxic methods for pest control, such as neem oil or insecticidal soap. Regularly clean out filters and containers to rid them of unwanted larvae, particularly if your systems are stored outside. Sticky traps can help monitor and reduce flying insects.

For diseases, ensure good air circulation and avoid overwatering to prevent fungal issues. Regularly inspect your plants for the subtlest signs of pests or diseases and act swiftly to address any problems. Integrated Pest Management (IPM) strategies, which combine biological, cultural, and mechanical methods, are highly effective in indoor setups.

Lighting is a key factor in indoor gardening, as choosing suitable grow lights can make all the difference. LED grow lights are highly efficient, producing less heat and consuming less energy compared to other types. They are ideal for plants at all stages of growth. Fluorescent lights, such as T5 fixtures, are also effective, especially for seedlings and leafy greens.High-intensity discharge (HID) lights, including Metal Halide (MH) and High-Pressure Sodium (HPS), are powerful and suitable for flowering stages but generate more heat and consume more energy. You can source these lights from most gardening stores or online retailers, with prices ranging from $50 for basic LED setups to $300 for advanced HID systems.

You'll need to maintain hydroponic and aquaponic systems through regular upkeep and troubleshooting. For hydroponics, ensure your nutrient solution is regularly refreshed and the pH is balanced. Clean the system components to prevent algae and mineral buildup. In aquaponics, good water quality is maintained by monitoring fish health and ensuring bio-filters are functioning correctly. Some of the common issues include nutrient imbalances, algae growth, and equipment failures. You can address these by regularly inspecting your system, cleaning and making necessary adjustments, and replacing faulty components promptly. Keeping a maintenance log can help track changes and improvements over time.

Don't forget to go download some additional tips on both hydroponics and aquaponics at the end of this book as part of your free bonuses.

11.6 SECURING URBAN APARTMENTS AGAINST THREATS

Urban living brings its own unique security challenges. Apartments are often more accessible to intruders, making it crucial to assess and address potential vulnerabilities, including burglaries and home invasions. Start by evaluating the security of your apartment building and individual unit, paying particular attention to weak points like old locks and unsecured windows.Many apartment buildings have shared entryways, which can be a security risk if not properly monitored. You'll want to ensure that any entry points are secure and consider additional measures like better lighting and surveillance.

One of the priorities in enhancing security is reinforcing entry points. High-security deadbolts and door reinforcement kits can significantly improve your apartment's defenses. Install deadbolts that are resistant to picking and bumping. Door reinforcement kits can strengthen the frame and hinges, making it harder for intruders to force entry. Securing windows is equally important. Use window locks, bars, or security film to prevent unauthorized access. Adding a door brace can provide an extra layer of protection against forced entry. These measures make it more hassle for intruders to break in, giving you and your family more peace of mind.

Implementing surveillance and alarm systems is another very effective way to enhance security. Choose security cameras suitable for both indoor and outdoor use. Cameras like the Arlo Pro4 are highly recommended for their affordability, ease of installation, and high-definition video quality. For motion detectors and smart alarms, consider the Ring Alarm system, which is known for its reliability and can be integrated with other smart home devices. Installing a smart doorbell camera, such as the Ring Video Doorbell, provides real-time alerts and allows you to monitor visitors through a mobile app. These systems help deter intruders and provide valuable evidence in the case of a security breach.

Creating a safe room within an urban apartment can be challenging, but it's possible with careful planning. Select a spot that is easily accessible yet secure, such as an interior closet or bathroom. Reinforce the safe room with additional security measures, including a sturdy door with multiple locks and reinforced walls. Stock the room with emergency supplies like non-perishable food, water, medical kits, and communication devices. Refer to Chapter 10 Safe Spaces for instructions on construction and reinforcement. Having a safe room provides a secure refuge during emergencies and will give you added confidence in ensuring your family's safety.

11.7 REAL-WORLD EXAMPLES OF URBAN SURVIVAL

Surviving civil unrest in urban areas is not something we want to think about but is something we really have to consider, as it requires foresight and decisive action. Last year, there were riots in my city. Luckily I had left the area prior to things kicking off. But I want to tell you about one of the shopkeepers in the city center who had planned ahead, and his foresight saved his store and his livelihood. He fortified the front with reinforced doors, windows, shutters, and security cameras, and this helped his property remain secure when the chaos and looting erupted. The costs were significant, but the investment paid off when he was able to protect his whole inventory of products and continue operations while others faced severe losses. This shows just how important it is to take proactive steps to safeguard your assets.

Another compelling story is about a family who decided to shelter in place during a citywide curfew. They had a well-stocked pantry and enough emergency supplies to ride out the situation safely. Their pantry was filled with non-perishable foods and plenty of water, and they had essential medical supplies just in case. This kept them going through the crisis until everything calmed down. The family's readiness really highlights the importance of having a solid emergency plan and supplies. Staying calm and self-sufficient during the unrest allowed them to feel safe and secure.

Urban prepping for pandemics has also proven crucial, as seen during the COVID-19 pandemic. One apartment dweller took the initiative to create a small indoor garden to supplement their food supply. Using vertical gardening techniques, they maximised their limited space and grew a variety of vegetables and herbs. This not only provided fresh produce but also reduced trips to crowded grocery stores. Simple solutions can offer more food security and self-sufficiency even in small spaces.

Navigating power outages in the city presents unique challenges. One city resident shared their experience: they had set up a reliable backup power system in advance that kept essential appliances running, including their refrigerator and medical devices, using a portable generator and solar panels to maintain power during a long blackout. They were able to stay comfortable and stress-free during the interruption.

Another family relied on a community support network during a major blackout. They shared resources and information with neighbors to stay informed and safe. This included pooling together for news updates and sharing extra supplies like candles and blankets. Their collective effort shows how the community working together and mutual aid during emergencies benefits everyone.

Urban gardening and self-sufficiency practices have transformed lives. A rooftop gardener in the heart of the city can successfully grow vegetables and herbs using containers and raised beds. Their garden not only provides fresh produce but can also become a therapeutic and productive hobby.

A community that establishes a shared garden and composting system offers another inspiring example, utilizing a vacant lot to create a communal garden where members can grow their own food and compost organic waste. This can foster a sense of community togetherness and shared purpose. It demonstrates how collective efforts can promote sustainable living practices and provide more food security for everyone involved.

11.8 RESOURCE VAULT

To further learn how to maximize the potential of your small space, here are several sources that can provide more insights.

The Urban Farmer by Curtis Stone is a fantastic book that offers practical advice on profitable small-scale farming. Stone's experience of urban agriculture translates into actionable steps that you can implement in your own small internal space, balcony, or rooftop.

For those looking to dive deeper into container gardening, Container Gardening for All Seasons by Barbara Wise is an excellent guide. It offers seasonal tips and plant recommendations tailored for container setups. This book helps you make the most of your containers, ensuring year-round productivity.

Websites like Epic Gardening (epicgardening.com) are treasure troves of information, offering articles, tutorials, and videos on various aspects of urban gardening, from soil care to pest control, and they update their content regularly to keep you informed about new products and techniques.

When it comes to equipment, the Click and Grow Smart Garden 9 is a top recommendation. This indoor garden comes with everything you need in the pack and uses smart soil and automated watering to take the pressure off you, making it significantly easier to grow herbs and small vegetables indoors.

For aquaponics enthusiasts, The Aquaponic Farmer by AdrianSouthern and Whelm King provides a comprehensive guide on setting up and maintaining an aquaponic system. The book covers everything from design principles to troubleshooting, making it a really valuable go-to for information.

Another recommended tool is the Mars Hydro TS 1000 LED Grow Light. This light is highly efficient, providing the full spectrum needed for plant growth while consuming less energy. It's ideal for both hydroponic and soil-based indoor gardens, making sure your plants receive adequate light even in low-light environments.

Don't forget Rob Bob's Aquaponics and Backyard Farm, whose YouTube channel has a wealth of knowledge and demonstrations on how to put together DIY aquaponic setups with lots of insider tips.

Urban survival is about making the most of what you have and turning challenges into opportunities. With these materials and tools, you'll be well-equipped to transform your urban living space into a productive environment, whether you're growing your own food or securing your home.

So far, we have covered several vital shelter-in-place strategies. In the next section of the book, we'll explore critical elements to prepare for any unexpected evacuation scenario, starting with how to assemble and prepare your bug-out bag. We'll also discuss a number of essential wilderness survival skills.

Your Feedback Matters

If the Prepper's Survival Bible has been useful to you, I'd be really grateful if you could leave a short review on Amazon or Goodreads. Your feedback doesn't just support this book, it helps other readers find the resource, decide if it's right for them and gain skills that might one day make a real difference.

Reviews are one of the best ways to strengthen the prepping community. Even a few words can help guide someone new, encourage others to start, and keep these skills alive.

In our community, we rely on each other's experiences and insights to stay ahead of the curve. Your review could be the catalyst that helps another prepper make a crucial decision or learn a new skill that one day could literally save someone's life.

Thank you for helping others get prepared.

Let's keep the community strong... together.

https://authorhelp.uk?azr=1917501013

Your thoughts matter to us

Thank you!

EVACUATION: BUGGING OUT

BUG-OUT BAGS

Living in the Wild
Extinctionis the rule.
Survival is the exception.

CARL SAGAN

Image courtesy of Anthony Acosta

The power goes out, communication lines are down, and you have minutes to evacuate. In such a critical moment, having a well-prepared bug-out bag can be the difference between chaos and calm. A BOB is your survival kit designed to help you "bug out" of a serious situation. Think of your BOB as your personal superhero sidekick, poised to swoop in and save the day when life throws you a curveball. Pack your BOB with everything you need to last at least 72 hours as an insurance policy against those "uh-oh" moments. Whether you're facing a natural disaster, civil unrest, a surprise pandemic, or any other significant crisis, your BOB will be your grab-and-go lifesaver.

12.1 BUG-OUT BAG ESSENTIALS

A BOB is designed with one main goal in mind: to help you survive and stay functional when you need to leave your home quickly. The ideal size for your bag is one you can carry comfortably, yet large enough to hold the essentials. Having multiple bags for different family members ensures everyone has what they need. Also, consider having backups and duplicates as much as you can without overloading, so you don't get caught out.

- Water and Food:Start with water and purification tablets and a portable water purifier. Non-perishable, high-energy food and snacks are next. Think protein bars, nuts, and dried fruits.
- First Aid Kit: Include your comprehensive first aid kit with essential medications. This should cover basic injuries and common ailments.

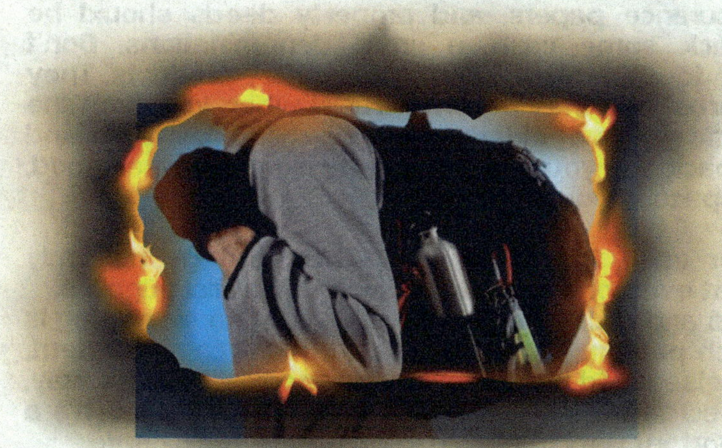

Image courtesy of James Carroll

Layered Clothing: Clothing is crucual.

- Start with a base layer of Merino wool or synthetic fabrics like polyester, which wick moisture and prevent heat loss. Avoid cotton as it retains moisture.
- The mid-layer should be fleece, down, or synthetic insulation like PrimaLoft for warmth.
- The outer thermal layer should be waterproof and windproof, using fabrics like Gore-Tex. Brands like 5.11 Tactical and Helikon-Tex are popular among the prepper community.

- A beanie, insulated gloves, and waterproof pants complete the setup
- Survival Footwear: Reliable, waterproof, and insulated boots are essential. Brands such as Salomon, Merrell, and Danner are highly recommended. Pair them with moisture wicking socks and consider gaiters for added protection.
- Tools and Equipment:
- A multifunctional tool is indispensable. The Leatherman Wave+ is a favorite.
- For knives, the Morakniv Garberg, priced around $82, offers excellent durability and edge retention. Sharpening methods include portable sharpeners like the Fallkniven DC4.
- Prevent rust with regular oiling and store knives in a dry place.
- Legal considerations vary, so check local laws
- Carry flashlights and radios with extra batteries, and have a variety of fire-starters and paracords.
- A whistle can attract attention if needed.
- Gadgets:Opt for devices that don't rely on electricity: Crank-up radios, solar chargers, and solar or wind-up flashlights or headlamps are practical choices.
- Portable power banks: Keep your devices charged.
- If there is a risk that you may have to wing it in the wild, check the Wilderness Survival: Extended Gear Check list in the Bug-Out Bag downloadable bonus at the end of this book. You'll also see a 72-hour Checklist for a basic BOB.

Having a prepped bug-out bag means you're ready for the unexpected, providing peace of mind and a safety net for you and your family.

12.2 CUSTOMIZING YOUR BUG-OUTBAG

When it comes to customizing, it's not really a cookie-cutter approach. Your unique circumstances, like your location, family needs, and the specific risks in your area, play a big role in what you should pack.

If you're living in a hot, arid region, hydration and sun protection are your best friends. Load up on water, throw in some electrolyte tablets, and choose lightweight, breathable clothing. And don't skimp on a wide-brimmed hat and a high-SPF sunscreen; those are essentials you'll be glad to have.

On the flip side, if you're in a colder, wetter environment, you'll want to prioritize waterproof gear, reliable gloves, and good thermal insulation. A few hand warmers and a compact sleeping bag rated for lower temperatures can add that extra layer of comfort.

Plus, don't underestimate the power of personal touches. Items that bring you comfort or familiarity can make a world of difference in tough situations.

Important documents, identification, insurance papers, and property deeds should be waterproofed and securely stored. Pack some cash in small denominations. Don't overlook personal hygiene essentials like toothbrushes, soap, and sanitary products. They might feel like small potatoes now, but trust me, they become crucial when you're on the move. For families, think about the specific needs of elderly relatives, children, infants, and pets. For example, a well-fitted small pack for your dog can carry some of their food, water, and medical supplies, freeing up space in your main bag.

Consider having multiple bags: one for each family member. This strategy not only lightens the load but also guarantees that everyone is equipped with essential supplies. You can share the heavier gear, like tents and cooking equipment, so that no single bag is overloaded. Assign specific items to each person; for example, one might carry the tent, another the cooking gear, and someone else could handle the larger medical kit. This way, if one bag goes missing, you're not left scrambling without all your critical supplies. It's a team effort, and each person has an important part to play in keeping the family safe.

12.3 MAINTENANCE AND UPKEEP

Maintaining and updating your bug-out bag is essential. If you reach for it and find expired food or broken gear, it's already too late. Regularly scheduled maintenance ensures everything is in working order. Check expiration dates on non-perishable foods and medical items every six months. Replace anything nearing its end.Seasonal updates are also essential. Swap out gear based on changing weather conditions. For instance, replace lightweight clothing with thermal layers as winter approaches. Keep a checklist to track these updates and set reminders on your phone or calendar.

Maximizing space in your bug-out bag isn't just about fitting everything in; it's about making the bag manageable and easy to carry. Use a layering system to pack your bag: Place the heaviest items, like water and tools, closer to your back and at the bottom.This keeps the weight centered and helps you maintain balance. Lighter items, such as clothing, should be on top. Compression sacks are handy for reducing the size of bulky items like jackets or sleeping bags. They allow you to pack more efficiently and create additional space in your bag.

Multi-use items will save space and reduce weight. A survival knife, for example, can be used for cutting,self-defense, and even starting a fire. Vacuum-sealed bags are excellent for packing clothing or food. They significantly reduce space while keeping contents dry and protected. Organize your gear into smaller pouches or bags based on categories like food, first aid, and tools. This modular packing system not only maximizes space but also makes it easier to find what you need quickly. Regularly review and repack your bag to ensure it remains efficient and set for any situation.

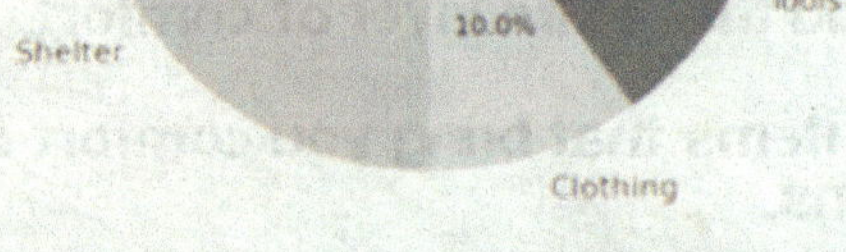

Graph 5: Bug-Out Bag Weight Distribution
This pie chart illustrates how to distribute the weight of essential items in a bug-out bag, ensuring an even load for greater mobility during evacuation.
Based on general survival and emergency preparedness guidelines.

12.4 RESOURCE VAULT

The list of items wecould pack in a BOB is endless, so I'm giving you several invaluable tools and sources that can help you refine your approach, starting with books.

The Prepper's Guide to Bug-Out Bags by Jim Cobb is a great read. It covers everything from selecting the right gear to packing efficiently.

Another excellent resource is Build the Perfect Bug-Out Bag by Creek Stewart, which offers practical advice and real-world examples.

For other online sources, the Ready.gov website provides comprehensive guides on emergency planning and keeping ahead, including detailed checklists for bug-out bags.

Another useful site is MIRA Safety, which offers expert reviews and recommendations on protective gear and equipment.

In terms of tools, the Leatherman Wave+ stands out as a versatile multi-tool. Its robust design and multiple functionalities make it reliable in any survival situation. For knives, the Morakniv Garberg is known for its durability and edge retention. It's a solid investment at around $82.

The Fallkniven DC4 is a compact and effective sharpening tool to keep blades in mint condition.

For lighting, the Fenix PD35 Tactical Flashlight is highly recommended. It's durable, waterproof, and has a long battery life. Pair it with rechargeable batteries and a solar charger like the Anker PowerPort Solar for a sustainable power source.

The Baofeng UV-5R is an excellent handheld radio for communication, offering a range of frequencies and a durable build.

These resources and tools will help you build a BOB that's not just functional but optimized for any scenario. Regularly updating your knowledge and gear will ensure you're always equipped to handle the unexpected.

WILDERNESS SURVIVAL SKILLS

You're deep in the forest, miles away from any signs of civilization, and nightfall is approaching. The temperature drops, and you're far from the comfort of your home. This is where wilderness survival skills become invaluable. Knowing how to navigate, find water, and build shelter can make the difference between a safe, albeit challenging night, and a dire situation. In the wild, there'sno room service. Knowing how to tie knots, start a fire, and not eat the wrong berries is just part of the charm.

13.1 PREPARING FOR WILDERNESS LIVING

> By failing to prepare, you are preparing to fail.
>
> BENJAMIN FRANKLIN

Your vehicle is your lifeline when it comes to evacuations. Ensuring it's in peak condition for any situation is crucial. Start with the obvious: routine maintenance, including regular oil changes, tire rotations, and brake checks, are non-negotiable. A well-maintained vehicle is way less likely to break down when you need it most. Always keep your gas tank at least half full to avoid running out of fuel when you least expect it.

Stock your car with essential tools and supplies. A spare tire, wheel wrench, and jack are basics, but consider adding a more comprehensive tool kit. Items like duct tape, a multipurpose tool, jumper cables, and an emergency car kit (including flares, blankets, a flashlight, and extra batteries) are vital. Don't forget a car cell phone charger, drinking water, and non-perishable food like protein bars. Keep a first aid kit and fire extinguisher close at hand. In colder climates, in addition to snow chains for your tires, add an ice scraper, shovel, and sand or kitty litter for traction. Regularly check and refresh these supplies to ensure they're always in good condition.

Fuel storage and management are paramount for long-term readiness. Store extra fuel safely in approved containers and use fuel stabilizers to maintain its quality. Plan your vehicle load for efficient evacuations. Remember to pack heavier items low down and towards the center to maintain balance and control; the same idea applies to your car. Place heavier items lower down and as close to the middle of your vehicle as possible. In emergencies, every detail counts.

You can access your free bonus download with a vehicle preparedness checklist at the end of this book. Don't forget to take advantage of all your bonuses!

Getting ready for life in the wilderness means picking up some essential skills and finding water tops the list. While humans can manage without food for weeks, we can only go a few days without water, so it's crucial to know where to look. Valleys or low-lying areas are your best bets as water flows downhill, so it could lead you to water sources like rivers and streams.

Look for dense plants or even changes in the sky's color; these can be signs of nearby water. You can also learn a lot from wildlife; game trails and bird flight patterns often lead to water sources. Just remember to purify any water you find before drinking to steer clear of contaminants. For a deeper dive into purification methods, check out Chapter 5, and the Water Filtration Cheat Sheet downloadable bonus.

In the wilderness, your compass is more reliable than your smartphone's GPS. The bonus? It never runs out of battery, and it won't reroute you through a lake. Learn to use maps, magnetic compasses, and protractors. Natural indicators like the position of the sun and stars can help you find your way. The North Star, for example, will give you a reliable marker for true north. Use the sunrise and sunset for directional cues. Pay attention to landmarks and create mental maps to keep your bearings.

Waterproofing your gear is critical in the wilderness. Use waterproof bags and containers to protect essential items. Dry bags are excellent for keeping clothes and food dry. For smaller items, you can use ziplock bags. Apply waterproof coatings to gear and use tarps to shield your camp from rain. Remember, staying dry is crucial for maintaining body temperature and comfort.

Equip yourself with gadgets that don't rely on electricity. A hand-crank radio, like the Midland ER310, provides weather alerts and communication without the need for batteries. Solar chargers can keep your devices powered when you're off the grid. A flashlight with a hand-crank or solar charging option ensures you won't be left in the dark.

13.2 SURVIVAL KNOTS

Tying a knot might seem simple, but trust me, when it's the difference between a secured tarp and waking up soaking wet, you'll be glad you practiced. Even a basic knowledge of knots is enough to help you secure your shelter, fish, and hunt. Here are some of the handiest knots with links for you to see how to tie them, to ensure your plans don't unravel at the worst possible moment.

SCAN ME!

The Square Knot is perfectfor joining two ropes together.Start by holding the rope so the two tail ends face away from your body. Cross the right tail end over the left tail end and pass it under. Wrap what is now the left tail-end up and over the right rope tail, then pass it under and through. Finally, pull tight.

https://www.animatedknots.com/square-knot

SCAN ME!

The Bowline Knot creates a secure loop at the end of a rope. Form a small loop, pass the tail end through it, around the standing part, and back over the loop created.

https://www.animatedknots.com/bowline-knot

SCAN ME!

The Taut-Line Hitch (aka Rolling Hitch or Magnus Hitch) is adjustable, perfect for tensioning tent guy lines. Start with a normal knot by wrapping the rope around the standing part, then wrap it around and through again, then finish with a half hitch by bringing the end away from the existing loop or knot, and pass the end through the new loop you have just created once and pull tight.

https://www.animatedknots.com/rolling-hitch-knot

SCAN ME!

The Trucker's Hitch provides significant tension for securing loads or pulling tarps tight. Create a loop in the middle of the rope, pass the end through a fixed point, back through the loop, and pull tight, finishing with two half hitches, or you can add a halter style hitch or slip knot for added stability.

https://www.animatedknots.com/truckers-hitch-knot

SCAN ME!

The Improved Clinch Knot secures fishing lines to hooks. Thread the line through the hook eye, wrap it around the standing line five times, pass the end through the loop near the hook eye, and then back through the loop formed. Pull to firmly secure your knot.

https://www.animatedknots.com/improved-clinch-knot

SCAN ME!

The Snell Knot is useful for securing a hook to the line when the line is passed through the hook eye twice before wrapping it around the shank of the hook. Pass your line through the hook eye. Create a loop on the hook end of the hook. Hold in place, and with the tail, wrap around the shank and loop 8 to 10 times, leaving the end part of the loop free. Pass the tail end through that remaining loop near the hook end. Dampen the knot area and pull tight. Trim any excess tail off.

https://www.animatedknots.com/snell-knot

For emergency use, the Slip Knot and Figure Eight Knot are essential:

SCAN ME!

The Slip Knot is quick-release, making it ideal for situations where you need to untie a rope rapidly. Create a loop, put your hand through the loop you created and grab another part of the rope. Pull your hand back out of the loop, while still holding the rope. Your other hand will need to secure the rope so you can do this. Pull tight and you shouldhave a slip knot that you can adjust to be larger or smaller by pulling on one side of the rope.

https://www.animatedknots.com/slip-knot

SCAN ME!

The Figure Eight Knot creates a secure loop at the end of a rope, useful for climbing and rescue. Form a loop, twist once away from the tail of the rope, then pass the end through the loop. You can then pass the end through or around an object and then feed the rope back through the knot, following the existing rope's journey so it looks like you have double ropes going through the knot. This provides a very strong stable knot.

https://www.animatedknots.com/figure-8-knot

You may remember some of these knots from the Boy Scouts (now Scouting America), but regular practice is crucial, especially since doing them under pressure can be tough. Use different rope types to get a feel for how each knot behaves. In high-pressure situations, muscle memory can be your best ally.

13.3 ADVANCED KNOTS

While the basics are essential, mastering more advanced knots can give you an edge in challenging situations. In situations where you may need to deal with rescue work, heavy loads, or challenging outdoor scenarios, these advanced knots are worth adding to your skill set. Follow the links for demonstrations you can practice with.

Alpine Butterfly Knot: The Alpine Butterfly Knot creates a secure loop in the middle of a rope that holds strong under tension. It's ideal for climbing, rescuing, or isolating damaged sections of rope. Pro Tip: Great for when you need to bear weight on both ends of the rope.

SCAN ME!

Alpine Butterfly Knot: The Alpine Butterfly Knot creates a secure loop in the middle of a rope that holds strong under tension. It's ideal for climbing, rescuing, or isolating damaged sections of rope. Pro Tip: Great for when you need to bear weight on both ends of the rope.

https://www.animatedknots.com/alpine-butterfly-loop-knot

SCAN ME!

Bowline on a Bight: This knot forms two secure loops, making it useful for rescue operations or quickly creating a seat harness. Pro Tip: Handy for creating a makeshift harness or extra loops on the fly.

https://www.animatedknots.com/bowline-on-a-bight-knot

SCAN ME!

Prusik Knot: This friction hitch allows you to climb or descend a rope safely. It grips tightly when weight is applied but slides when not under load. Pro Tip: Perfect for ascending or descending ropes when you don't have climbing gear.

https://www.animatedknots.com/prusik-knot

SCAN ME!

Double Fisherman's Bend: The Double Fisherman's Knot securely ties two ropes together, often used in climbing or rescue. It's stronger and more reliable than the regular fisherman's knot. Use it when you need to join two ropes or create a loop that won't slip.

https://www.animatedknots.com/double-fishermans-bend-knot

These more advanced knots are practical, versatile, and will give you essential skills you'll need for any situation. Even if you don't need them every day, mastering them will add valuable tools to your survival toolkit.

13.4 RESOURCE VAULT

These more advanced knots are practical, versatile, and will give you essential skills you'll need for any situation. Even if you don't need them every day, mastering them will add valuable tools to your survival toolkit.

When it comes to wilderness living, knowledge is your most valuable tool. Books, websites, and gear can provide the insights and practical advice you need to thrive in the wild. Let's look at some recommended reading and tools that can expand your understanding and equip you with the necessary skills for off-grid living.

First, consider The SAS Survival Handbook by John 'Lofty' Wiseman. This book is a treasure trove of survival information. Written by a former SAS soldier, it covers everything from building shelters to finding food and water. The detailed illustrations and step-by-step instructions are invaluable, whether you're just starting out or looking to refine your skills.

For those interested in advanced navigation techniques, Be Expert with Map and Compass by Bjorn Kjellstrom offers valuable information. This book demystifies all aspects of orienteering, making it accessible to beginners while offering more depth for seasoned navigators.

It covers the basics of reading topographic maps and using a compass effectively. The exercises included help cement your learning of the concepts, so you can navigate confidently without relying on GPS.

Understanding how to find your way using natural indicators and traditional tools can be essential in wilderness survival. Websites like Scouting Magazine offer valuable information. Their article,Finding Water in the Wilderness, provides great practical advice on finding and purifying water sources.

Reading online articles will help you stay updated on the latest techniques and innovations in survival skills. Websites also offer community forums where you can ask questions, share experiences, and learn from others. Engaging with these communities can provide you with real-time advice and support.

When it comes to survival equipment, the LifeStraw Personal Water Filter really stands out. This compact, portable device allows you to drink directly from streams and lakes, filtering out harmful bacteria and parasites. It's a lifesaver in situations where clean water is scarce. It is lightweight and easy to carry and pack in your bug-out bag. Pair it with a portable solar charger like the Anker PowerPort Solar for a reliable energy source when you're off the grid. Keeping your devices charged can be critical.

For practicing knots, the website Animated Knots by Grog offers clear, step-by-step animations. Watching these animations can help you understand the intricacies of each knot, making it easier to practice and master them. The site covers a wide range of knots, from basic to complex, ensuring you have the knowledge you need for any situation. Practicing these knots regularly will build your muscle memory so you can tie them quickly and efficiently under pressure.

Finally, the Midland ER310 Emergency Crank Weather AM/FM Radio is an excellent addition to your survival gear. This radio provides access to NOAA weather alerts, which are is vital for staying informed about changing weather conditions. Its hand-crank and solar charging capabilities mean you'll always have power, even in remote locations. The builtin flashlight and ultrasonic dog whistle add extra layers of utility, making it a versatile tool for any emergency scenario.

Equipping yourself with the right knowledge and tools supports your ability to survive and thrive in the wilderness. This provides a solid foundation for boosting your readiness skills. As we move forward, we'll get into the intricacies of foraging, a skill that strengthens your self-sufficiency and ensures you have the means to sustain yourself and your family during extended emergencies.

FORAGING FOR WILD EDIBLES

Image courtesy of Michael Burrows

Foraging is like a journey into the heart of nature. Regular foragers often speak of the profound connection they feel with the land as they identify edible plants and mushrooms. This hands-on experience nurtures a deep appreciation for ecosystems and the delicate balance of life within them. For many, foraging brings a sense of empowerment and self-sufficiency, reminding them of our ancestors' ways of living off the land. It's not merely about the harvest; it's about embracing the adventure, learning the rhythms of nature, and relishing the satisfaction that comes from providing for oneself and loved ones.

14.1 IDENTIFYING AND HARVESTING PLANTS IN THE WILD

Foraging is an essential skill for anyone serious about living sustainably off-grid. It enhances food security and variety, reduces dependency on cultivated crops, and fosters a unique bond with nature. Proper identification is crucial to avoid adverse effects from poisonous plants. I strongly recommend taking a foraging course to give you hands-on experience and confidence to know what you're doing. You might already know the basics, but advanced knowledge, like recognizing subtle differences between similar species, is a necessity. For example, understanding the nuances between wild garlic and its toxic look-alike, Lily of the Valley, could save your life.

Foraging laws and regulations vary by location. Always check local guidelines before venturing out. Foraging on private land without permission is illegal and unethical. Public lands come with their own set of rules to keep things balanced. The golden rule here is simple: take only what you need and leave enough behind for wildlife and plants to bounce back. This practice not only preserves the ecosystem but ensures future foraging opportunities. For instance, when foraging dandelions, pick only a few leaves and flowers from each plant to allow it to thrive and propagate.

Identifying common edible plants starts with knowing their key characteristics. Safe-to-eat plants often have distinct leaves, flowers, and stems. Of course, dandelions, for instance, are easily recognizable with their toothed leaves and bright yellow flowers and are entirely edible, from roots to flowers. Wild garlic, with its broad leaves and white star-like flowers, emits a distinct garlic smell, making it easy to identify. Nettles, another valuable wild edible, are also easily recognizable and have serrated leaves and tiny stinging hairs. They're best harvested with gloves and cooked to neutralize the sting.

Images courtesy of WesAbrams, MarenWischnewski and dulezidar

Habitat and growth patterns are equally important. Plants like wild garlicthrive in damp, shady woodland areas. Understanding where plants like to grow is key. Spring and early summer are prime times for foraging many greens, while berries are abundant in late summer. Understanding each plant's growth habits ensures you harvest at peak freshness and nutritional value.For example, harvesting wild garlic involvesgently pulling the bulbs without damaging the plant, ensuring it can regrow.

Image courtesy of Snrdjanpav

Foraging Must-Haves

Field Guide: You'll need to carry a reliable field guide with photos for comparison. Errors are not an option with foraging.

Tools: Use sharp tools for clean cuts.

Containers:Collect in breathable containers like mesh bags or baskets.

Notebook and pencils: Document plant locations and characteristics for future reference and draw sketches.

Foraging connects you with your environment in a deeply satisfying way. Each plant you identify and harvest adds to your knowledge and self-reliance. It's a skill that, once honed and practiced, becomes second nature.

14.2 PRESERVING AND USING FORAGED FOODS

Once you've successfully foraged, preserving those wild edibles ensures you can enjoy them long after the harvest. Drying is one of the simplest and oldest methods. Lay cleaned and dried plants on a tray or hang them in bundles in a well-ventilated area away from direct sunlight. Once dried, store them in airtight containers in a cool, dark spot. This method works well for herbs such as mint and chamomile. Picture the end of the day, sitting down with a tea made from herbs you foraged yourself. This is so easy to do, and far more satisfying than store-bought.

If you have access, freezing is another effective method. Briefly blanch your plants in boiling water, then transfer them to a container of iced water. Drain them well and place them in freezer safe containers or bags. This technique is especially useful for greens, berries, and certain mushrooms. For instance, freshly picked wild strawberries can be frozen and later used in smoothies or desserts, retaining their nutrients and flavor.

Fermenting, on the other hand, adds a tangy twist to your wild edibles. Make a brine by mixing salt into water, and then submerge the cleaned plants into a fermentation vessel. After a few weeks at room temperature, transfer them to the refrigerator for longer storage. You can find out more about this process in Chapter 3 on Food Preparedness.

Including foraged foods in your meals can be both simple and creative. Foraged greens like dandelion leaves make excellent additions to salads and nettles to soups; for example, a hearty nettle soup, rich in vitamins and minerals, or a fresh dandelion salad with a tangy vinaigrette. The nutritional benefits are immense. Many wild edibles are packed with vitamins, minerals, and antioxidants, offering health advantages that cultivated crops might lack. Combining foraged foods with your garden produce creates balanced and nutritious meals. For example, wild garlic can enhance the flavor of homegrown tomatoes in a delicious, savory pasta dish.

14.3 SAFETY AND ETHICALCONSIDERATIONS

Foraging can be incredibly rewarding, but it comes with responsibilities. Safety should always be your top priority. Misidentifying plants can have serious consequences. For instance, mistaking a toxic plant like hemlock for an edible one could lead to severe poisoning.

One of the best ways to ensure you're foraging safely is to take a local foraging course. Learning from knowledgeable mentors can help you avoid the pitfalls from inexperience. Field guides are invaluable tools, but hands-on guidance from an expert is irreplaceable. Joining localg roups can also foster a sense of community as group members often share tips on sustainable harvesting and help newcomers learn the ropes. They provide a supportive forum where you can ask questions and get immediate feedback.

Avoiding toxic plants is obviously non-negotiable. Learn to recognize common poisonous species and steer well clear.Hemlock, with its smooth stems and small white flowers, is often mistaken for wild carrot. Amanita mushrooms, some of the most toxic fungi, can be confused with edible varieties.

Foraging in safe areas is another key aspect. Avoid sites near roads or industrial areas where plants may be contaminated with pollutants. Instead, seek out untouched environments like forests or meadows far from human activity.

Using field guides and other resources can significantly enhance your foraging skills in the wild once you have already participated in foraging courses. Look for guides with detailed images and descriptions to help you accurately identify plants.

Respecting nature is fundamental to ethical foraging. Over harvesting can devastate local plant populations and disrupt ecosystems. Only ever take what you need, leaving plenty for wildlife and plant regeneration. This practice ensures that the plants will continue to thrive, supporting both nature and future foragers.

14.4 RESOURCE VAULT

To stay safe when foraging, have the right information at your fingertips. Books offer a wealth of knowledge, and one highly recommended read is The Forager's Harvest by Samuel Thayer, which goes into the identification and use of various wild edibles with detailed illustrations and practical advice, making it an invaluable guide for both beginners and seasoned foragers.

Another excellent resource is Edible Wild Plants: Wild Foods from Dirt to Plate by John Kallas. It provides in-depth information on foraging techniques and recipes, helping you make the most of your wild harvests.

Websites are also fantastic for up-to-date information and community support. Eat The Weeds (eattheweeds.com) is a comprehensive resource for identifying and using wild plants. The site includes videos, articles, and a forum where you can share experiences with other foragers and ask questions.

Another valuable online resource if you are in the UK is Wild Food UK (wildfooduk.com), which offers detailed plant profiles, foraging tips, and information on local foraging courses. These websites are great for continuous learning and staying connected with the foraging community.

When it comes to foraging, access to the right equipment can enhance your experience. A sturdy, reliable field knife like the Morakniv Companion is essential for safely and efficiently harvesting plants.

A good pair of gloves, such as the Atlas Nitrile Garden Gloves, will protect your hands from thorns and stings while foraging. Additionally, a breathable mesh bag is perfect for collecting your finds without causing them to wilt or spoil.

Lastly, consider using a smartphone app like PlantSnap to help with plant identification. This app allows you to take a photo of a plant and get instant information about its species, edibility, and uses. Combining these resources and tools will significantly enhance your foraging skills and confidence for times when you need them.

Of course, bear in mind that if SHTF and the internet goes down, you'll have to rely on your field guides, notes, and prior expertise. This is why practicing foraging now as a way of life will serve you well in the midst of a crisis.

Foraging connects you to nature in a profound way, providing not just food but also knowledge and a sense of self-sufficiency, which is priceless in uncertain times. Next, we will explore some fishing and hunting techniques that could help you and your loved ones to survive and thrive in the wild.

HUNTING AND TRAPPING TECHNIQUES

15.1 SURVIVAL FISHING

SPOOLS OF FISHING LINE, HOOKS AND SINKERS. IMAGE COURTESY OF GAL_ART

You're deep in the wilderness, far from the nearest town, and your stomach is rumbling. The sun is setting, and you realize you need to find food… and fast. Fishing canbe a real lifesaver in situations like this, offering not just a meal but also a satisfying sense of achievement. So, let's look into the essentials of survival fishing, from pinpointing the best spots to catching and preparing your meal.

Fishing stands out as one of the most dependable ways to secure food in the wild. It takes patience, a bit of know-how, and the right gear. Start with the basics: fishing line, hooks, and lightweight nets. These essentials are versatile and can be adapted for various fishing techniques. A small spool of fishing line and a handful of hooks hardly take up any space but can truly make a difference when you need to eat.

Finding the right water source is key. It's not just about spotting rivers, streams, or lakes; you want to target areas where fish are likely to hang out. Look for shallow, slow-moving waters where fish tend to gather. In colder climates, ice fishing can be a viable option.

Select a safe spot on the ice, drill a hole, and use a small hook and line to catch fish beneath the surface. Fly fishing, on the other hand, involves casting a lightweight lure that mimics insects, making it ideal for catching trout and other freshwater species in flowing streams.

Improvising fishing tools can be a lifesaver when you don't have traditional gear. Crafting a fishing spear from a sturdy branch by sharpening the end can be effective. Aim below the fish to account for the water's refraction. You can also create fish traps using natural materials. For instance, make a basket trap from flexible branches, shaping them with a funnel entrance to catch fish as they swim in. Using your hands, rocks, or sharpened sticks can also work in a pinch. Techniques like hand fishing, known as "noodling" in some regions, involve grabbing fish directly from the water. Arrange rocks in a 'U' shape facing the flow of water so the current directs fish into the enclosure, making them easier to grab or spear. Incidentally, if you have seen the movie The Revenant, you'll have seen this exact method in action.

Different environments host various fish species. In freshwater, you might encounter trout, bass, or catfish. Saltwater areas could yield mackerel, snapper, or flounder. Identifying these species helps tailor your approach for better success. Bait selection is another critical factor. Insects, worms, and natural lures attract fish effectively. You should find plenty of these under rocks and debris in moist soil. Timing also matters; early morning and late evening are prime fishing times when fish are most active.

Tailoring bait to lure target species requires understanding their preferences. For example, trout are attracted to insects, while catfish prefer strong-scented baits like chicken liver.

Use these six tips to optimize your bait:

1. Match the hatch: use local insects
2. Keep it fresh
3. Adjust size based on fish type
4. Add scent for attraction
5. Use bright colors in murky water
6. Experiment with different baits to see what works best

Once you've caught your fish, it's essential to dispatch, process, and preserve it properly. Quickly dispatch the fish to minimize suffering, then clean and gut it to prevent spoilage. Rinse the fish thoroughly and remove the innards. To preserve, consider smoking, drying, or salting, if you plan to store it for later use.

Fishing in the wild isn't just about catching fish; we must do so ethically and sustainably. Routinely practice catch-and-release when possible and avoid overfishing to ensure the ecosystem remains balanced.

Safety should always come first: purify your catch by cooking it thoroughly to avoid foodborne illnesses. Make sure you heat fish to an internal temperature of 145°F or above (63°C). Proper handling and care ensure that your wild-caught meal is both safe and satisfying.

15.2 HUNTING AND TRAPPING

Hunting and trapping are essential skills to have in your survival toolkit, especially when other food options are running low. It's all about knowing the land and understanding the animals that call it home. First and foremost, ethics matter. Always respect wildlife and adhere to local laws. Hunting should really be a last resort: approach it with care and consideration for the animals, and only when you absolutely have to. Become familiar with the local wildlife by observing common game and their behaviors. For example, deer tend to be on the move during the early morning and late evening, while rabbits and squirrels are often easier to spot at dawn.

In a specific region, sustainable hunting practices might involve targeting overpopulated species to maintain ecological balance. For example, a local community might focus on hunting wild boar to prevent them from damaging crops and habitats. This not only provides food but also helps manage wildlife populations responsibly.

Let's move on to the tools and techniques. Bow hunting requires selecting the right bow and arrows. Longbows and compound bows are popular choices. Practice regularly to improve accuracy and efficiency. Firearms, like rifles and shotguns, offer more power but come with greater responsibility. Choose a rifle for long-range hunting or a shotgun for close encounters. Always follow safety tips: treat every firearm as loaded, never point at anything you don't intend to shoot, and keep your finger away from the trigger until ready to fire.

Tracking and stalking require patience and keen observation. Learn to read animal tracks and understand behavior patterns. Look for signs like broken branches, droppings, and tracks. Move slowly and quietly, staying downwind to avoid detection. Use natural cover to get closer for a clear shot or to set a trap.

Building and setting traps can be highly effective for catching small game, although it is a crueler way to trap your prey. Deadfall traps use weight to crush prey. Construct a simple deadfall with a heavy rock and a trigger mechanism. Snare traps, made from wire or cordage, can catch rabbits and squirrels. Position them along animal trails where they're likely to pass. Pit traps are useful for larger game. Dig a deep hole and camouflage it with branches and leaves, ensuring it's hidden from the animal's view. Set and check traps during early morning or late evening when animals are most active.

Handling trapped prey requires skill and care. Dispatch the animal quickly and humanely. We want to minimize suffering at all costs. Process the game meat by field dressing, which involves skinning and gutting the animal. Use a sharp knife to make clean cuts. Remove the innards and thoroughly rinse the cavity to prevent bacterial growth. For long-term storage, refer to methods like drying, smoking, and salting discussed in our food preparedness chapter. These techniques ensure the meat remains safe and edible for extended periods.

15.3 RESOURCE VAULT

When it comes to honing your hunting and trapping skills, having the best tools at your disposal can make all the difference. Whether you're an experienced survivalist or just starting, here are some recommendations to provide valuable insights and practical advice to enhance your knowledge and skills.

For books, start with Bushcraft 101: A Field Guide to the Art of Wilderness Survival by Dave Canterbury. This book breaks down essential outdoor skills, including hunting and trapping, with straightforward instructions and illustrations.

Another excellent resource is The Trapper's Bible: Traps, Snares & Path guards by Dale Martin. This book offers detailed explanations and diagrams for a wide variety of traps. It's a comprehensive guide that will expand your knowledge and improve your trapping success.

Websites like Outdoor Life (outdoorlife.com) feature articles, videos, and tutorials on survival fishing, hunting, trapping, and include community forums where you can swap experiences and advice with fellow survivalists.

Another great site is the Missouri Department of Conservation (mdc.mo.gov), which offers detailed guidelines on ethical hunting practices and wildlife management. Their hunter ethics section is particularly valuable for understanding the legal and moral aspects of hunting.

For survival fishing, consider a compact fishing kit like the Best Glide ASE Survival Fishing Kit. It's lightweight to easily fit in your bug-out bag, and includes hooks, lines, and lures.

When it comes to trapping, a set of Victor Conibear traps will serve you well for catching small to medium game. These traps are durable and effective and are ideal for setting up multiple traps in different locations.

For practical field dressing and game processing, the Outdoor Edge Wild Pak is an all-in-one kit that includes knives, a saw, and other necessary tools. This set is designed for efficiency and ease of use, ensuring you can process your catch quickly and safely.

Additionally, the Leatherman Signal multi-tool offers versatility with its pliers, wire cutters, and other features, making it a handy companion for any survival situation.

Having covered the basics of hunting, trapping, and fishing, it's time to shift gears. In the next chapter, we'll learn all about building temporary shelters so you'll have a cozy spot to call home no matter where your adventures take you.

BUILDING TEMPORARY SHELTERS

Picture yourself in the wilderness, nightfall approaching, with no permanent shelter in sight. The temperature drops, and you need to actfast to protect yourself and your loved ones from the elements. In such moments, you'll really want to know how to build a temporary shelter. Let's talk about the essentials of staying warm and dry in any situation.

16.1 STAYING WARM AND DRY

Building a shelter in the wild is a bit like playing with nature's LEGO set. Only this time,the stakes are a little higher than stepping on a stray brick. Choosing the right type of temporary shelter is crucial. Factors like weather conditions, portability, space requirements, and ease of setup are significant. In cold climates, you'll need materials that provide insulation, like heavy-duty tarps or thermal blankets. Conversely, in hot weather, breathable materials to allow airflow are essential. Lightweight and easily transportable shelters are ideal, especially if you need to move quickly. Consider options like emergency bivvies or tube tents, which are compact yet effective. Space is another consideration; ensure your shelter can accommodate all occupants and necessary supplies. Lastly, prioritize shelters that are quick to assemble and disassemble. Speed can be a lifesaver in emergencies.

Setting up tents involves various techniques. Dome tents, tunnel tents,and pop-up tents each have their advantages.

- Dome tents are lightweight and easy to set up but can be unstable in high winds.
- Tunnel tents offer more room but may collapse under extreme conditions.
- Pop-up tents are convenient but often less durable

.

Begin by selecting a safe, level site free of debris and away from water flow paths. Anchor the tent securely using stakes or guy lines, especially in windy conditions. Ensure proper ventilation to reduce condensation and maintain airflow. Packing essentials include a ground tarp, extra stakes, and a repair kit for quick fixes. For experienced survivalists, practice setting up your tent in various conditions to ensure speed and efficiency.

Using tarps for shelter provides versatility. Lean-to, A-frame, and flat roof configurations are among the most effective. Choose a sturdy tarp, ideally made of coated nylon or silnylon, and use ropes, stakes, or trees to secure it. This is where your knowledge on knots will be really useful. For a lean-to, tie one end of the tarp to a tree at chest height and stake the opposite end to the ground. The A-frame requires a ridge line between two trees or trekking poles. Drape the tarp over the ridge line and stake the corners to create a tent-like structure. Waterproofing is critical: use additional tarps or natural materials like leaves and branches to enhance protection against rain and wind. Always carry lightweight cord and ultralight stakes so you can adapt to various environments.

IMAGES COURTESY OF SSHEPARD AND JAMES MAHAN

Constructing improvised shelters using found materials is a skill everyone should master. Lean-to shelters and debris huts are simple yet effective. Use branches for the frame and cover with leaves or debris for insulation. In snowy conditions, quinzhees and snow caves offer excellent protection. Digging a T-shaped entrance and hollowing out the interior creates a warm, insulated space.

When it comes to bushcraft shelters, use natural materials like logs and foliage. Urban environments require more creativity. Cardboard and plastic sheeting can form handy makeshift shelters. The key to all these shelters is selecting the right location.

In fact, selecting the best location for your shelter is as important as the shelter itself. Avoid low-lying areas prone to flooding, stay near water sources and firewood but ensure you're not too close to potential hazards, and position your shelter to protect against prevailing winds and harsh weather.

A well-chosen site not only provides safety but also enhances the effectiveness of your shelter, keeping you warm and dry when you really need it.

16.2 WASTE DISPOSAL

This is a subject we don't like to discuss, but we don't have the luxury of ignoring it if we find ourselves in the wild, trying to survive. Managing waste in off-grid or emergency situations is not just an unpleasant discussion topic; your health can depend on it. If improperly managed, waste can potentially spread nasty contaminants and illnesses to you and your family. If you're in a remote area, designate a specific spot for waste, away from living areas and water supplies.

To avoid the spread of diseases and pollution, use containers with tight lids to store waste temporarily before disposal. When burying organic waste to promote natural decomposition, make sure it's at least six inches deep to keep animals away. For human waste, follow the same depth guideline and cover it with soil to minimize odors and reduce the attraction of flies.

Constructing a basic composting toilet or pit latrine system can significantly improve sanitation and reduce the risk of contamination, once you set it up far away from water sources.

- A composting toilet system can be built using a sturdy wooden frame (for sitting on) and a bucket. Line the bucket with a biodegradable bag (you'll need a steady supply).
- For a more permanent pit latrine setup, dig a pit at least six feet deep and build a simple structure above it.
- Again, ensure the pit is far from any water sources to prevent contamination. When the pit is full, you can pick a new spot to dig and set up.

Using natural materials for effective odor control can make a significant difference. Covering waste with sawdust, straw, leaves, lime, or wood ash can really control odors and accelerate decomposition.

Charcoal can also be used to absorb odors and moisture. If you're in an area with pine trees, pine needles work well as a natural deodorizer. Incorporate these materials into your composting toilet or pit latrine system to maintain a more pleasant environment. For added benefit, create a simple air vent or exhaust pipe to allow odors to escape while keeping your toilet shelter well-ventilated.

Essential hygiene practices are vital to maintaining health in off-grid conditions. Always wash your hands with soap and water after handling waste.

If water is scarce, use hand sanitizer with at least 60% alcohol. If you find yourself with no supplies, soapwort, yucca, and clematis can act as soap substitutes.

Establish a routine for cleaning and disinfecting your living area, especially around waste disposal sites. Regularly clean tools and containers used for waste management with a solution of bleach and water where possible. Remind all family members about the importance of hygiene to prevent the spread of germs and illnesses.

16.3 RESOURCE VAULT

To truly excel in building temporary shelters and managing waste disposal effectively, having the insightful guides at your fingertips is crucial. Here are some recommendations with valuable insights and practical advice to enhance your skills and knowledge.

For those interested in deepening their shelter-building expertise, The Ultimate Survival Manual by Rich Johnson is an invaluable resource, covering a wide range of survival techniques, including detailed instructions for various types of shelters.

You can find step-by-step guides and tips for creating effective shelters in diverse environments using natural materials in Bushcraft 101 by Dave Canterbury.

When it comes to managing waste disposal, The Humanure Handbook by Joseph Jenkins is a must-read for longer-term wilderness stays. This book explores sustainable waste management, such as composting human waste safely and effectively. It provides comprehensive guidelines and practical advice for setting up and maintaining composting systems, essential for anyone living off-grid.

The Composting Toilet System Book by David Del Porto and Carol Steinfeld offers detailed instructions and case studies on various composting toilet designs. This book is particularly useful for those looking to implement eco-friendly sanitation solutions in their off-grid setups.

Online platforms can also be incredibly helpful. Websites like Survival Sherpa (survivalsherpa.com) offer a wealth of information on shelter building, waste management, and other survival skills, featuring articles, tutorials, and videos that cover a wide range of topics, making it a valuable resource for continuous learning.

Another excellent online resource is The Prepared (theprepared.com),which provides in-depth guides and reviews on survival gear, including tents, tarps, and waste disposal systems to help you make informed decisions when selecting the best tools and equipment for your needs.

For shelter construction, a high-quality survival knife like the Morakniv Companion is indispensable and perfect for cutting branches, handling materials, and other essential tasks.

A folding saw, such as the Silky Gomboy, is also a valuable addition to your toolkit. It allows you to quickly and efficiently cut through wood, making it easier to construct sturdy shelters.

For waste management, consider investing in a portable composting toilet like the Nature's Head Self-Contained Composting Toilet. This user-friendly system is designed for off-grid living and provides an effective and eco-friendly solution for managing waste.

Having the right information is vital for mastering the skills needed to build temporary shelters and manage waste effectively. I hope these quality books, online resources, and suggested tools help you to be as well-prepared as possible so you can face any emergency or off-grid situation.

Our next chapter tackles the critical aspects of fire-starting, ensuring you have the knowledge and skills to stay warm, cook food (especially if you don't have the luxury of a portable stove), and signal for help in any situation.

FIRE-STARTING

Image courtesy of Secoura

You and your family are huddled around a makeshift campsite, the temperature dropping as the sun sets. You're miles away from home comforts, and the only thing standing between you and the cold night is your ability to start a fire. Fire is more than just a source of warmth; it's a beacon of hope, a means to cook food, and a way to keep predators at bay. Mastering fire-starting techniques is essential for anyone serious about off-grid living and survival.

17.1 FIRE-STARTING TECHNIQUES

Starting a fire is a basic skill that every survivalist needs to master. Traditional methods like using flint and steel or a bow drill are tried and true. Flint on steel is one of the easiest primitive fire-starting methods. All you need is a piece of flint and a carbon steel striker. Striking the flint against the steel creates sparks, which ignite a dry tinder bundle. The bow drill, though more complicated, requires less exertion once built. Create friction by spinning a spindle against a fire board using a bow and string. This method can be more reliable in damp conditions where other methods might fail.

Modern methods, while often more convenient, should never be solely relied upon. Lighters are the most straightforward option, but they can run out of fuel or malfunction. Waterproof matches are another excellent choice, especially in wet environments. Fire starters, like magnesium rods or the highly recommended Pyro Putty, provide a reliable and weatherproof fire-starting option. Pyro Putty can burn for up to 8 minutes, giving you ample time to get a fire going even in adverse conditions.

Images courtesy of Marco_Piunti, Jose Martinez, and Prill

Have you ever tried starting a fire in the rain? That's when you really prove you're not just a weekend camper but a full-fledged survivalist. Starting a fire in difficult conditions requires special techniques. When it's wet, finding dry tinder and kindling is challenging. Look for materials under logs, rocks, or inside dead trees. You can also use alternative fire starters like batteries and steel wool, or chemical fire starters. These methods can provide reliable ignition even in adverse conditions.

When it comes to fire safety, it's crucial to manage and extinguish fires properly to prevent wildfires. Always clear the area around your fire of flammable materials. Keep water, sand, or dirt nearby to extinguish the fire if necessary. Never leave a fire unattended. Digging a small trench around your fire can help contain it, especially in dry conditions.

Remember, safety should always be your top priority. Having multiple fire-starting tools available at all times is a fundamental preparedness principle. Keep lighters, waterproof matches, and a fire starter in your bug-out bag, vehicle, and home. This redundancy ensures that you're never caught without a means tostart a fire, no matter the situation. Whether you're using traditional methods or modern conveniences, the ability to start a fire is a cornerstone of survival.

17.2 MAINTAINING AND UTILIZING FIRE

Fire is a critical survival tool. Beyond the obvious need for warmth, fire provides the means to cook food, purify water, and signal for help. In colder climates, a well-maintained fire can be the difference between life and death. Even in milder conditions, fire offers comfort and security.

The warmth of a fire is such a relief after a day of foraging, along with the satisfaction of a hot meal cooked over its open flames. But keeping a fire going efficiently requires some knowledge and practice.

Different fire types serve specific purposes. The teepee fire is excellent for quickly getting a flame going. Arrange your kindling in a cone shape, with the tinder in the center. This structure allows air to flow freely, helping the fire to catch and build heat rapidly. Once you have a strong flame, you can transition to other types of fires.

For a longer-lasting fire, the log cabin setup is ideal. Stack your logs in a square formation, alternating directions with each layer.This structure burns steadily and provides a stable base for cooking. For a more directional fire, useful in windy conditions, the lean-to fire is your go-to. Lay a large log or branch on the ground and lean smaller sticks against it, creating a wall that shields the fire from the wind while funneling air through the fuel.

Using fire for signaling is another crucial skill. Smoke signals can alert rescuers to your location. Creating a dense smoke requires green leaves or damp wood. Adding these to a hot fire will produce thick, white smoke visible from a distance.

Starting a fire begins with choosing the right tinder and kindling. Tinder includes highly flammable materials like birch bark, dry grass, or cotton balls soaked in petroleum jelly. These catch fire easily from a spark. Kindling, such as small sticks, twigs, and dry leaves, helps sustain the flame until larger logs ignite.

Wind direction plays a significant role in fire maintenance. Position your fire so the wind carries flames and heat towards your fuel. This ensures a hotter,more efficient burn. In windy conditions, use natural barriers like rocks or build a windbreak to protect your fire from being extinguished.

17.3 ADVANCED FIRE-STARTING AND CAMPFIRE BUILDING

Building a reflective wall is a practical way to maximize the heat from your fire and protect it from the wind. To construct one, gather logs, rocks, or a combination of natural materials. Position the wall about 3-4 feet from your fire. This distance ensures that the heat is effectively reflected back towards you and your camp while maintaining safety. Arrange the materials in a semi-circle around the fire, with the open side facing your camp. This setup not only helps in retaining heat but also provides a windbreak, ensuring your fire burns steadily.

Creating afire pit with a windbreak is essential for a controlled and efficient fire. Start by selecting a suitable site away from overhanging branches and flammable materials. Dig a pit about one foot deep and three to four feet in diameter. Use the displaced soil to build a low wall around the pit. Stack rocks or use earth to create a windbreak on the side facing the prevailing wind. Ensure there's enough airflow; too much wind protection can smother the fire. This setup keeps the fire contained and makes it easier to manage, especially in windy conditions.

Building cooking rigs over the fire enhances your outdoor cooking experience. A simple yet effective method is to construct a tripod. Use three sturdy branches, lash them together at the top, and spread the legs out over your fire. Hang pots or kettles from the center using hooks or chains. For more versatility, create a crossbar setup by placing a sturdy branch across two forked sticks driven into the ground. Attach hooks to the crossbar for hanging multiple items. Adjust the height of your cooking pots by moving the hooks along the crossbar or using chains of varying lengths. This allows precise control over cooking temperatures.

Images courtesy of Andriiafanasiev and Zakharchenko

Creating a long-lasting fire involves specific techniques. Layering logs is a method where you place larger logs at the bottom and lighter ones on top. This setup ensures the fire burns steadily without frequent tending. For stable, long-lasting flames, consider the log cabin or pyramid structures. The log cabin method, with its interlocked logs, provides a stable base, while the pyramid structure ensures a slow, even burn. To keep a fire smoldering overnight, use the banking method: cover the embers with ash and a few larger logs. This conserves heat and makes rekindling easier in the morning.

Understanding and controlling smoke patterns is crucial. Position your fire so the smoke is carried away from your camp.Use natural barriers like rocks or hills to direct the smoke.

Remember that in survival situations, smoke can be an effective way to communicate a distress signal. Create dense smoke by adding green leaves or damp wood to a hot

fire. This produces thick, white smoke visible from a distance. Always be mindful of your surroundings and ensure your smoke signals are clear and intentional.

Fire Safety Recap

Always clear the area around your fire of flammable materials. Before lighting the fire, prepare water, sand, or dirt nearby for quick extinguishing in the event of the fire spreading. Never leave a fire unattended. For added safety, dig trenches around the fire to prevent spreading, especially in dry conditions, and use the displaced dirt to create a low barrier. In emergencies, create firebreaks by clearing vegetation around the fire or using dirt and sand to smother flames. These methods ensure you can quickly control and extinguish the fire if it gets out of hand.

17.4 RESOURCE VAULT

When it comes to mastering fire-starting and campfire building, having the right tools at your fingertips can make a significant difference.

The Ultimate Survival Manual by Rich Johnson includes detailed instructions on how to build fires in various conditions.

The Bug Out Bag Guide website offers a wealth of information on primitive fire-making methods. It breaks down techniques like flint and steel, fire plows, and the bow drill, providing step-by-step guides and videos.

Another valuable site is Outdoor Life, which offers reviews of the best fire starters. Their expert testing criteria include burn time, wind performance, and ease of ignition, ensuring you choose the most reliable tools to suit your needs.

For tools, the Exotac fireROD is a top recommendation. This ferrorod is waterproof, has an ergonomic handle, and the rod is replaceable, making it a long-lasting addition to your gear.

Pyro Putty is another excellent choice, known for its weatherproof properties and long burn time, making it ideal for survival situations. For a DIY option, consider using a combination of dryer lint, an egg carton, and candle wax. This homemade fire starter is lightweight, reuses waste, and has a respectable burn time.

A portable stove will get you through for a while too, such as the Primus Omnifuel which runs off canister gas or liquid fuel, or MSR WhisperLite International which are versatile and work off a variety of liquid fuels. Or you could opt for the more lightweight Vargo Titanium Alcohol Stove.

Incorporating these resources into your toolkit will enhance your fire-starting capabilities. Whether you're relying on a trusted book, accessing expert advice online,or using top-tier tools, you'll be well-prepared.

Next, we'll explore emergency communication plans, ensuring you stay connected and informed when it matters most.

EMERGENCY COMMUNICATION PLANS

Luck is what happens when preparation meets opportunity.

SENECA THE YOUNGER

You're cut off from the world during a rising flood. Traditional communication channels fail, and you have no way to contact your loved ones or get vital information. The thought of this scenario shows us why it's so important to have reliable communication networks in place. When my family and I were caught in a severe storm, we lost all cell service and internet access. It was a harrowing experience, not knowing what was going on or how long it would last. A robust emergency communication plan would have really helped.

18.1 SETTING UP RELIABLE COMMUNICATION NETWORKS

Choosing the right communication devices is crucial for maintaining contact during emergencies. Hand-crank radios are a staple, providing access to news alerts and weather updates without having to rely on batteries. They're indispensable in keeping you informed when the grid is down. For areas with poor cell coverage, satellite phones offer a lifeline. Devices like the Garmin inReach Mini 2 is compact, reliable, and provide two-way messaging over the Iridium satellite network, ensuring you can send and receive messages even in the most remote locations.

IMAGE COURTESY OF LAURA D YOUNG

Handheld two-way radios are another essential tool. They offer excellent range, reliability, and ease of use. Popular models like the Midland GXT1000VP4 provide up to 36 miles of range, making them ideal for family communication in rural areas. These radios are cost effective, have long battery life, and come with features like weather alerts and S.O.S. signals. Basic operation involves setting channels and adhering to radio etiquette, such as using call signs and emergency channels.

Ham radios are a step up, offering extended range and capabilities. They require a license, but the benefits are substantial. With a Technician License, you can access local and North American frequencies, while a General License opens up worldwide communication. The Ham community is supportive, offering a wealth of knowledge and assistance. In emergencies, when other systems fail, ham radios can be a critical communication tool.

CB radios are another option, especially for vehicles and local networks. They provide medium-range communication and are user-friendly.

Setting up a wireless mesh network (WMN) in your neighborhood by connecting wireless access point nodes (WAP) ensures a robust local communication system. This network can operate independently of the internet, providing a reliable backup.

Investing in solar chargers and portable power banks for your cell phones and other devices is wise. Keeping them fully charged and in designated spots like your bug-out bag, car and safe room ensures you have power during outages. Comparing the range, cost, and reliability of these tools shows that each has its strengths, and a combination of them will provide comprehensive coverage.

Backup communication methods are essential when primary systems fail.

Social media platforms can be invaluable for emergency updates if internet coverage is available. Community bulletin boards also play a crucial role in disseminating information. For example, during the Camp Fire in Paradise, California, bulletin boards became indispensable for sharing updates and coordinating relief efforts. Local radio stations can also provide timely information and should be included in your communication plan.

Emergency signaling methods are fundamental for anyone serious about disaster preparedness. Using signaling mirrors for long-distance communication and effective hand signals for group coordination can save lives. Creating and recognizing universal distress symbols, such as SOS, ensures that you and your family can signal for help even when traditional communication is impossible. Emergency audio signals, like whistles or air horns, are also important to keep on hand. Improvising signals with available materials, such as creating a signal fire or using flashlight patterns, can be effective in complex environments like urban areas or dense wilderness.

By integrating these communication strategies and tools into your emergency plan, you ensure that you and your family stay connected and informed during any crisis.

18.2 STAYING INFORMED: USING EMERGENCY ALERTS AND WEATHER TRACKERS

Beware of false knowledge; it is more dangerousthan ignorance

GEORGE BERNARD SHAW

Staying informed through emergency alert systems is vital for your safety. Real-time updates can make the difference between life and death during a crisis. What if a sudden wildfire encroached on your community? Emergency alerts can provide the critical information you need to evacuate in time. These systems offer weather updates, evacuation orders, and public safety notices, ensuring you're always aware of evolving situations.

To set up emergency alert systems on your devices, start by enabling Wireless Emergency Alerts (WEAs) on your smartphone. These alerts are government-issued and cover a range of emergencies, from natural disasters to Amber alerts. Most modern smartphones have WEAs enabled by default, but it's wise to check your settings to ensure you're covered. Additionally, subscribing to local emergency notification services can provide more localized and timely updates. Many municipalities offer text or email alerts for residents, keeping you informed about issues specific to your area.

As I write this, I'm watching live news reports on a Category 4 Hurricane about to hit Florida. Weather tracking tools are essential for staying ahead of adverse conditions. Popular apps like AccuWeather, The Weather Channel, and NOAA Weather Radar Live provide detailed forecasts, radar images, and severe weather alerts. AccuWeather, for instance, offers comprehensive planning features, including allergy management and safe driving tips. The Weather Channel provides hourly and daily forecasts along with robust video content, while Weather Underground offers smart forecasts and storm tracking. Using these apps, you can interpret weather radar and forecast data to monitor approaching storms and plan accordingly. For example, seeing a severe thunderstorm approaching can give you time to secure your home and gather necessary supplies.

Creating a personal alert network enhances your readiness. Forming a neighborhood watch group allows for the rapid sharing of updates and key information. Establish communication protocols with nearby friends and family to ensure everyone is on the same page. Setting up a group text message system will facilitate the quick dissemination of crucial information. For instance, during a power outage, you can quickly inform your network about the status of your home, share resources, or coordinate a meeting point. This personal network acts as a safety net, providing support and information when official channels may be overwhelmed or delayed.

18.3 RESOURCE VAULT

Here is a summary of some practical advice and recommendations that will equip you with the knowledge and gear needed for effective communication during emergencies. Books are a great starting point for anyone serious about emergency communication. Emergency Radio Communications by Michael Martens is a must-read. It provides a deep dive into using ham radios, covering everything from basic setup to advanced communication techniques. This book is particularly valuable for those looking to get their ham radio license, offering step-by-step instructions and practice questions.

Another essential read is The Prepper's Communication Handbook by Jim Cobb. This book offers practical advice on setting up various communication systems, including CB radios, ham radios, and satellite phones. Cobb's experience in the field provides real world examples that make the concepts easy to grasp.

The ARRL (American Radio Relay League) website is an excellent resource for those interested in ham radio. It provides information on getting licensed, finding local classes, and accessing practice exams. The site also features forums where you can connect with other enthusiasts and seek advice.

Another valuable online resource on communication is Ready.gov, which offers guidelines on setting up emergency communication plans, including tips for using social media and local networks. The site provides templates and checklists that you can customize to fit your needs, ensuring you have a comprehensive plan in place.

The Garmin inReach Mini 2 stands out as an exceptional and reliable satellite communicator. Its compact size and robust satellite network make it ideal for remote areas. You can send and receive messages, track your location, and even trigger SOS alerts.

For those looking for a more budget-friendly option, the ACR Bivy Stick offers global coverage and on-device basic controls. It's perfect for occasional use and emergencies.

Handheld two-way radios like the Midland GXT1000VP4 are also indispensable. They offer an excellent range, multiple channels, and weather alerts, making them versatile for various scenarios.

Standard portable and solar power banks are also a must-have to have access to reliable backup power for your cell phones and radios.

As we wrap up this chapter, it's clear that having the right communication technology and knowledge is essential for staying connected during emergencies. These resources and tools will help you build a robust communication plan to keep you and your family informed.

We have now covered two of the three sections in our book: Sheltering-In-Place and Evacuation. Next, in our third and final section, we'll explore some of the long-term aspects of survival and recovery after a disaster. First, we'll zone in more deeply on the critical aspects of preparing the family, creating effective communication plans and ensuring everyone knows their role during a crisis. Then we'll cover some important aspects of financial and legal steps, along with other key components for staying ahead of whatever comes your way.

THE LONG HAUL PREPPING AND RECOVERY

PREPARING YOUR FAMILY

What if you're at work, your kids are at school, and your partner or spouse is running errands when suddenly, a major earthquake strikes. The ground shakes violently, buildings crumble, and communication networks go down. In that moment of chaos, having a well-thought-out family communication plan becomes a lifeline. Knowing that everyone is safe and has a clear course of action can bring an immense sense of relief amidst the turmoil. This chapter focuses on how to ensure your family stays connected and safe and knows what to do during these kinds of emergencies.

19.1 FAMILY COMMUNICATION PLAN

Staying connected during an emergency is absolutely critical.Communication is key, and losing contact with family can cause panic, lead to poor decisions, and make an already bad situation worse. We've seen this happen in real life. During Hurricane Katrina, a lot of families got separated and couldn't reconnect, which only added to the stress and uncertainty.

Having a solid communication plan helps to coordinate rescue efforts and makes sure everyone gets the help they need as quickly as possible.

Family communication is a major tenet of disaster planning. Clear and structured communication ensures that every family member knows their role and list of duties during a crisis.It prevents misunderstandings and reduces panic.When it comes to kids, it's important to share information in a way that makes sense for their age. You want them to feel safe and informed without overwhelming them. For instance, if you're practicing or actually evacuating during a wildfire, calmly explaining what's happening can prevent confusion and fear. Let them know the plan and where they fit and reassure them that they'll be okay. This gives them a sense of security, both during the drill and the real thing.

When practicing, walk your kid through the steps, like grabbing their go-bags or where to meet. Make it a routine so that when a real emergency happens, they know exactly what to do, so they will feel more confident:

- Hold regular family meetings to review the communication plan so everyone's clear on it.
- Agree on how you'll stay in touch, whether it's phones, walkie-talkies, or other methods.
- Establish regular check-in times during emergencies to account for everyone. This practice ensures that if one method fails, there are backups in place.

- Regularly practicing these protocols helps reinforce them, making them second nature when a real crisis occurs.

Creating a family communication plan involves several key steps. Designate primary, secondary, and even tertiary contact methods, and write everything down on your communication plan. For instance, agree to use a family group chat on a secure messaging app for regular updates and pre-arranged meeting points. Walkie-talkies can serve as a reliable communication tool when cell service is down, and so on. Establish a communication schedule for regular check-ins and identify a central meeting point in case of separation. Additionally, create a list of emergency contacts and important phone numbers. Distributing printed contact lists to all family members provides a tangible backup in case digital devices fail. You'll find a useful Family Communication Template in your bonus downloads at the end of this book.

Maintaining clear and calm communication during stressful situations is essential. Using techniques such as actively listening and repeating back what the speaker has said, will help them feel understood, validated and heard, and will provide them with much needed empathy and support. Avoid placing blame and instead focus on problem-solving. This approach fosters a supportive and cooperative environment, even under pressure. Clear and calm communication can help navigate the chaos and maintain order.

Supporting family members emotionally during crises is so important. Encourage open expression of feelings and concerns, offering reassurance and maintaining a positive outlook. For example, create a family support circle where everyone can share their thoughts and feelings. This practice fosters a sense of unity and mutual support, which is essential for emotional resilience. In times of crisis, emotional support can make a significant difference in how well a family copes and recovers.

Family Communication Checklist

Here is a quick 'To Do' list to help get you started, but don't forget to download a full Family Communication Checklist in your bonuses at the end of this book.

- Schedule regular family meetings to discuss communication protocols.
- Designate primary and secondary contact methods(e.g. Cell phones/WhatsApp, walkie-talkies).
- Establish regular check-in times during emergencies.
- Create a family group chat on a secure messaging app for updates.
- Use walkie-talkies for communication when cell service is down.
- Develop a communication flowchart with primary, secondary, and tertiary methods.
- Identify and communicate a central meeting point in case of separation.
- Create and distribute printed contact lists with emergency contacts and important phone numbers.
- Practice active listening and problem-solving during discussions.
- Encourage open expression of feelings and concerns within the family.

19.2 RESOURCE VAULT

Here is some valuable reading material to help grow your prepper's bookshelf:

Another must-read is the Crisis Preparedness Handbook by Patricia Spigarelli Aston. It's packed with info on getting ready for just about anything: natural disasters, pandemics, civil unrest, you name it. The book's detailed checklists and step-by-step guides for creating emergency plans make it a fantastic resource.

For online guidance, websites can be a goldmine, such as Ready.gov, which has tons of free templates and tips on building emergency plans. You can tweak their resources to fit your family's needs, like setting up communication plans, shelter strategies, and evacuation routes.

The American Red Cross website is another one to bookmark. They offer great advice on making emergency plans, responding to disasters, and even keeping family members emotionally supported when things go sideways. They also have online courses in first aid and disaster preparedness, which are super useful if you want to level up your skills.

Additionally, consider investing in portable power banks like the Anker PowerCore 10000. These compact and lightweight devices can keep your communication tools charged during power outages, ensuring you stay connected with your family.

> A man who does not plan long ahead will find trouble at his door
>
> CONFUCIUS

FINANCIAL PREPAREDNESS

You've heard there's a storm barreling towards your town. You're at the grocery store, watching the shelves empty out as people grab whatever they can. It's a chaotic scene, and all you can think about is whether you have enough supplies at home. Now, think of how different this situation would be if you had a dedicated survival budget that you applied over time to get in whatever you needed well in advance so you are primed for just such an event. Instead of feeling frantic, you'd be ready and relaxed. Financial planning is just as important as having the right gear. It ensures you can sustain your planning efforts without sinking into debt.

20.1 BUDGETING FOR PREPPING: COST-EFFECTIVE STRATEGIES

Taking action without a plan can quickly become a financial nightmare. You might find yourself buying random items, only to realize later that you've spent a fortune on things you don't need. Systematic, incremental investments in long-term preparedness help you avoid this pitfall. Think of it like building a house. You wouldn't buy all the bricks at once without a blueprint. Gradually stockpiling supplies ensures you stay within your financial limits while steadily building up your resources.

When it comes to cost-effective response planning, it's all about prioritizing purchases based on immediate needs and long-term goals. Start with essentials like food, water, and medical supplies. Utilize sales, discounts, and bulk buying to save money. For example, purchasing non-perishable foods in bulk during seasonal sales can significantly cut costs. Keep an eye out for deals on items like canned goods, rice, and beans. These staples have long shelf lives and provide essential nutrients.

Exploring DIY solutions can save you a ton of money. Instead of spending a lot on an expensive water filtration system, why not make your own with simple materials like cheesecloth, (clean) sand, gravel, and charcoal? And not forgetting thrift stores and garage sales, where you can often score affordable supplies like bug-out bags, camping gear, tools, and clothing for a fraction of the price. By taking a smart approach, you can get set for anything without overstretching your budget.

20.2 FINANCIAL PLANNING FOR LONG-TERM SUSTAINABILITY

Living off-grid comes with its own set of financial challenges, so having a solid plan is crucial. It's not just about getting through the next storm; it's about thriving for the long haul. Investing in areas like renewable energy systems and sustainable agriculture requires thoughtful financial planning. If you're not careful, costs can quickly spiral. So, a strong financial strategy helps you use your resources wisely and keeps you from running into money troubles later on.

To create a response planning budget, get an overview of your income, expenses, and savings. Pay attention to your discretionary income... that's the money you have left after taking care of the essentials. From there, figure out a percentage to set aside for your prepping fund

Even saving just 5% of your monthly income can really add up over time. Using budgeting apps like Quicken, Simplifi or YNAB (You Need A Budget) can be super helpful. These tools make it so much easier to track your expenses and set financial goals, keeping you organized and on track with your budget.

Also consider setting up a second dedicated savings account just for emergencies, like for supplies and repairs. You don't have to start big; even small, regular contributions can make a difference. This way, you're not just securing your financial future but also keeping your survival planning efforts steady, and effective along with catering for life's little emergencies.

Don't forget to set long-term goals that align with your vision of sustainable living. If you're thinking about installing solar panels, consider both the initial costs and ongoing maintenance. This shows why it's smart to have that second emergency fund alongside your main prepping expenses account. It'll give you a financial safety net and help you manage unexpected costs without derailing your readiness plans.

20.3 INVESTING IN PRECIOUS METALS AND BARTER GOODS

Investing in precious metals is the smart move for anyone looking to hedge against economic instability and inflation. Historically, metals like gold and silver have held their value through financial crises. When fiat currencies devalue, gold often retains its worth, providing a stable investment. The liquidity and value retention of precious metals make them a reliable financial asset. However, it's essential to be aware of market volatility and the need for secure storage. For instance, during economic downturns, gold prices typically rise as investors seek safe havens.

IMAGE COURTESY OF THAD

When it comes to investing in precious metals, you have a few options. Bullion, coins, and bars are popular, with gold being seen as a stable investment and silver offering a more affordable option with higher returns. You can even explore buying through ETFs (Exchange-Traded Funds). Just be sure to check the credibility of dealers to avoid scams. Do your research and read independent reviews to make informed choices.

Safe storage solutions include home safes and bank deposit boxes. For beginners, buying silver coins can be an accessible entry point, while gold coins provide a more significant investment. Both offer a tangible asset that can be easily liquidated if needed.

Barter goods will be essential in a post-collapse economy. When traditional currency loses its value, items like food, medical supplies, and tools become the new currency. It's important to stockpile things that people universally need, as these will be valuable trade assets.

When choosing what to include in your barter stash, think about its value, shelf-life, and demand. Creating a diverse inventory by organizing items based on their usefulness and demand can help make the most of what you have. And remember to update and rotate your inventory regularly to keep things fresh. Items like alcohol, batteries, and hygiene products are also great for bartering.

Effective bartering requires establishing networks and honing negotiation skills. Connect with local communities to create a barter network.Make sure you know the value of your items and stick to fair trading practices. It's also a good idea to keep track of your trades and inventory so you can stay organized. Use barter exchange templates and negotiation tips to ensure you're getting the best value for your trades. This strategic approach to bartering can help you stretch your assets and ensure you're well-prepared for any situation.

20.4 RESOURCE VAULT

Nailing the financial side of readiness can be a bit tricky for some. If you're looking for guidance, The Prepper's Financial Guide by Jim Cobb is a solid resource. It's packed with practical advice on budgeting, investing, and building a financial strategy that focuses on sustainability and self-reliance.

Another helpful read is Surviving the Economic Collapse by Fernando 'Fertal' Aguirre. He shares firsthand experiences from Argentina's economic crisis, giving you valuable lessons on managing resources and staying financially resilient in tough times.

For more information, websites like Investopedia offer useful articles on investing in precious metals. Their Precious Metals Investing Guide for Beginners covers everything from the basics to more advanced strategies, helping you make informed choices about gold, silver, and beyond.

If you want to buy, sell, or store precious metals, the GoldMoney app can be really handy. It provides real-time market data, making it easier to make quick decisions.

For budgeting and financial planning, Forbes' list of the best budgeting apps we mentioned, such as Quicken, Simplifi (by Quicken) and YNAB (You Need A Budget) can help you track your expenses and set financial goals. These apps offer features such as personalized spending plans, investment tracking, and real-time updates, making it easier to manage your finances and stay on top of your budget.

Finally, keeping a detailed journal of your financial plans and progress can be really helpful. Regularly updating it keeps you focused on your goals and lets you adjust your strategies as needed.

Checklists are also invaluable. Don't forget to download your Barter Checklist in the bonuses at the end of this book for a checklist of high-demand barter items to stockpile. You can edit it to ensure you cover all essential categories like food, medical supplies, and tools relevant to your area.

By using these tools, you'll be well-equipped to navigate the financial challenges of survival and off-grid living, ensuring you and your family remain secure and resilient in the face of any crisis. In the next chapter, we'll explore the legal considerations essential for off-grid living, ensuring you stay compliant with local regulations while you build your sustainable lifestyle.

LEGAL CONSIDERATIONS

You've finally found that perfect piece of land to build your off-grid paradise. You can almost feel the excitement as you picture your new, self-sufficient life. But before you start digging, there's a vital step you can't overlook: learning your local property and zoning laws. These rules can either pave the way for your dream or throw up roadblocks that derail everything you've planned. It's not just about following the law; it's about dodging fines and legal headaches that could set you back significantly.

21.1 UNDERSTANDING ZONING LAWS AND REGULATIONS

If you're aiming for an off-grid lifestyle, zoning laws are essential to understand. They dictate what you can and can't do with your land. These regulations can vary quite a bit depending on where you are, and affect what you can build and what energy systems you can install, for instance. Urban areas often have stricter rules than suburbanor rural ones. Some regions might even have guidelines that encourage sustainable practices, like rainwater collection or solar panels, or a rural property might welcome a wind turbine.An urban spot might have limits because of noise or appearance concerns. So, take the time to dig into the local laws; there may or may not be opportunities that align perfectly with your off-grid vision.

Ignoring these regulations can lead to some serious consequences, like hefty fines, legal troubles, or even being forced to take down your structures. Some folks have been known to invest time and money into building their dream off-grid home, only to face a legal order to tear it down. It's a nightmare scenario no one wants to experience.

Take the Johnson family, who moved from the city to a rural area to pursue off-grid living. They quickly realized they couldn't install their planned wind turbine due to local zoning restrictions. Instead of abandoning their project, they applied for a zoning variance and worked closely with local officials, eventually gaining approval to proceed.

Researching local regulations is your first step. Visit your local government's website, where you can find zoning maps and ordinances outlining what you can and cannot do on your property.

Common zoning restrictions for off-grid living often include strict building codes, dictating the materials and methods you can use for construction. This means you'll need to submit detailed plans and have your construction approved at various stages. You could end up planning a solar energy setup only to find that local laws restrict panel installation on visible roof surfaces.

When it comes to building and modifying structures on your off-grid property, understanding building codes and permit requirements is essential. These codes ensure your structures are safe and meet health standards. For example, if you plan to construct a rainwater harvesting system, you'll need to obtain the necessary permits and ensure your system complies with health and safety regulations. This might include using specific materials or installing backflow preventers to protect the public water supply.

When diving into the world of property regulations, it's essential to be aware of zoning overlay districts. These are special areas that come with extra rules on top of the standard zoning laws. Understanding these districts can give you insights into what might be permissible in your neighborhood. For instance, if you find that your area has an overlay promoting eco-friendly developments, you might have more flexibility with your projects than you initially thought.

Don't overlook local agricultural programs; they can be a goldmine for urban gardeners and homesteaders. Many regions offer initiatives that support urban agriculture, which can lead to exemptions or special considerations for growing your own food right in the city. Getting involved in these programs not only benefits you but also contributes to your community's overall sustainability.

If you're navigating around land use regulations, community land trusts can be a great ally to help you understand local laws. Their ethos is about making the most of your land while building a sense of community, and they sometimes offer options for affordable housing or cooperative farming.

And speaking of rights, knowing what you're entitled to as a homesteader or farmer is super important. Some states have right-to-farm laws that protect you from nuisance lawsuits. This can be a huge relief if you plan on growing food or raising livestock.

Another smart move is getting involved in local government. Attend town meetings or community forums to stay in the loop. Plus, it's a chance to share your thoughts on issues that impact how you can use your land. Your voice matters, and being proactive can really help shape your community.

Lastly, don't skip the history of your potential property. Taking the time to dig into past ownership and any previous zoning variances can give you valuable insights that might make your plans easier to navigate. Knowing past variances or exceptions can be incredibly informative and may even support your current application process.

Building relationships with other homesteaders and local preparedness groups in your area can be invaluable. These connections can provide insider knowledge about local zoning practices relevant to you and successful strategies for navigating them. Plus, it's a great way to foster a supportive community.

Keep environmental regulations in mind, too. Understanding local laws (like those protecting wetlands) can open up possibilities for sustainable practices that comply with zoning regulations. This knowledge can help you align your off-grid plans with the community's environmental goals.

If you're interested in renewable energy, check for any local incentives for installations. Some areas offer programs that might give you more leeway with zoning restrictions for things like solar panels or wind turbines. Taking advantage of these programs can benefit you while also benefiting the environment.

Finally, don't hesitate to seek legal advice when necessary. Knowing when and how to consult with professionals who specialize in property law can be crucial, especially if you're facing challenges from local authorities. Having the right support can help you navigate any legal hurdles and ensure your off-grid lifestyle remains secure.

Other legal considerations include staying aware of local laws that can vary significantly. Foraging laws, for example, can dictate where and what you can forage. Some areas prohibit foraging on public lands or restrict the collection of certain plants.

Whether you own a gun or not, it's good to know the basics about firearm laws in your state. Understanding the rules helps everyone stay safe and avoid any legal hiccups.

Emergency protocols, such as mandatory evacuation orders, can also have legal implications, so make sure to be familiar with them.

21.2 RESOURCE VAULT

Navigating the legal landscape of off-grid living can be complex, but with good information, you can gain a clear understanding of what's required and how to stay compliant. Here are some valuable books, websites, and other materials that provide detailed guidance on various aspects of property and zoning laws, building codes, and other legal considerations.

One of the top recommended books is Navigating the Legalities of Building and Living OffGrid by Sarah Hastings, which offers an in-depth look at the challenges and solutions associated with off-grid living. It covers everything from researching local building codes to obtaining necessary permits. Hastings provides practical advice on how to approach local authorities and build positive relationships with them, which can be crucial when seeking variances or exemptions. The book also includes case studies that illustrate how others have successfully navigated zoning laws, making it a practical guide for anyone looking to go off-grid.

If you're trying to untangle the maze of zoning laws, The Complete Guide to Zoning by Dwight Merriam is a solid choice. It breaks down all the jargon into clear language, covering how zoning works, what to know about land use and tips for handling disputes. This book is especially handy if you're getting knee-deep into agricultural zoning, offering insights that can help homesteaders, preppers and anyone working towards self-sufficiency. Plus, it includes real-life examples and step-by-step guidance for dealing with variances and local zoning boards.

For those venturing into building codes and permits, check out Building Off the Grid: A Practical Guide to Living in Harmony with Nature by Gary Collins. He covers the basics of building codes: how to read them, what permits you need for different projects, and how to ensure your builds meet health and safety standards. If you're thinking about installing solar panels or wind turbines, this book has the information you need in order to navigate the legal requirements and best practices.

If you're feeling overwhelmed by zoning regulations, you're not alone, and there are plenty of online sources to help. Websites like LegalMatch are a treasure trove for digging into local zoning laws. They'll help you find zoning maps, consult with planning departments, and prepare for variance applications. They also explain common restrictions so you know what to expect.

Another useful site is Crest Real Estate, which walks you through the process of applying for zoning variances, from filling out the paperwork to getting the nod from your neighbors. They share practical tips to improve your chances, like attending zoning board meetings or chatting with a zoning lawyer.

And don't overlook the ZoningCheckapp, which lets you enter your address and quickly pulls up zoning regulations for your area. You'll find out about building height limits and what types of structures you can put up. It even connects you to local zoning boards and planning departments so you can get your questions answered.

If foraging is part of your strategy, knowing the local laws is key. Foraging Laws and Regulations by John Kallas is an excellent guide for navigating around what you can legally harvest in different areas. It covers which plants are fair game and any permissions you might need. This resource is especially valuable for foragers looking to gather wild edibles without stepping outside the legal lines.

Gun ownership is another critical area where legal knowledge is essential. The Gun Owner's Guide to State Laws by Alan Korwin provides a comprehensive overview of gun laws in every state. It covers all the basics of gun ownership.

By understanding the ins and outs of property laws and zoning regulations, you're laying a solid foundation for your off-grid journey. This groundwork not only helps you sidestep potential legal issues but also empowers you to make informed decisions about your living situation. Now that you've equipped yourself with this knowledge let's explore some real examples of individuals and families who've successfully navigated crises, demonstrating the resilience and resourcefulness that are vital in challenging times.

REAL-LIFE SURVIVAL STORIES

You're sitting in your living room, watching the news. The meteorologist's voice is urgent as he describes the approaching hurricane. You glance at your family, and the gravity of the situation hits you. Now, picture a different scenario: you're prepared. You have a plan, supplies, and a network of support.

In fact, events like Storm Amy in Europe, the devastating Ofunato wildfire in Japan, and the deadly Cebu earthquake in the Philippines in late 2025 show how quickly crises can unfold. No region is immune.

This chapter is about real people who faced unimaginable situations and lived to tell the tale. Their stories offer invaluable lessons for anyone serious about keeping ahead.

22.1 LESSONS FROM NATURAL DISASTER SURVIVORS

When Hurricane Katrina hit New Orleans in 2005, the devastation was unimaginable. One family, the Smiths, had managed to evacuate just before the storm. Clint Smith remembers the chaos of grabbing their well-stocked bug-outbags and leaving everything behind. Their quick action made all the difference, highlighting how crucial it is to be ready to go at a moment's notice. A bug-outbag packed with essentials can truly be a lifesaver.

On the flip side, during Hurricane Sandy in 2012, a New Jersey neighborhood showed how powerful community effort can be. Neighbors came together, sharing resources like generators, setting up a communal kitchen, and making sure everyone had access to clean water. Their strong sense of community became a lifeline, showing that sometimes, working together is just as important as individual prepping.

Earthquakes present a different set of challenges, as experienced by many Californians. With over 500 active faults in the state, the risk of a major quake is ever-present. One resident, John, took proactive steps to retrofit his home for seismic activity. He reinforced the cripple walls, anchored the house to the foundation, and secured heavy furniture to prevent injuries. These actions significantly reduced the risk of damage and injury when the quake hit, highlighting the importance of reinforcing your home against natural disasters. Simple steps like securing your heavy furniture and water heater can go a long way.

During the 2011 Japan earthquake, another family found their emergency communication plan invaluable. The earthquake and subsequent tsunami disrupted all traditional communication channels. However, the family had multiple methods of staying in touch, including satellite phones and pre-arranged meeting points.Relying solely on one mode of communication can leave you vulnerable when it fails. Having backups like satellite phones and designated meeting points ensures you can stay connected even when the grid goes down.

Wildfires are an increasing threat, especially in places like California. One couple learned this the hard way but took preemptive measures to create a defensible space around their home. They cleared vegetation, created firebreaks, and kept an up-to-date go-bag for quick evacuations. When a wildfire threatened their home, they were ready to evacuate quickly. This story underlines the importance of proactive measures in wildfire-prone areas. Clearing vegetation and creating firebreaks can protect your property and give the fire service a fighting chance.

Another family faced the terrifying reality of evacuating with pets during the California wildfires. They had a pet emergency kit primed, complete with food, water, medications, and comfort items for their animals. This preparation ensured that their pets were safe and calm during the evacuation. A pet emergency kit is essential for families with animals, ensuring that everyone, including pets, is taken care of in an emergency.

Floods are another disaster that can hit with little warning. During the severe flooding in Nebraska in March 2019, one homeowner had taken precautions by installing flood barriers and sump pumps. Along with some well-placed sandbags, these efforts kept the damage to a minimum. It's a great example of how small steps can make a big difference when a flood hits. Simple things like sandbags and a sump pump can be invaluable for protecting your home.

In another case, a community pulled together for a large-scale evacuation during the 2019 Midwest floods. They had developed a community emergency plan that included designated evacuation routes, shelters, and communication protocols. This planning ensured a smooth and safe evacuation for all residents. Developing community emergency plans can enhance the overall readiness of your neighborhood and ensure everyone knows what to do when disaster strikes.

The Ice Storm of 1998 in northeastern North America left people without power for weeks, right in the middle of a freezing winter. Many turned to wood stoves and propane heaters for warmth, while neighbors chipped in by sharing things like generators and extra food. In one community, the way everyone worked together to clear ice and get basic utilities back up really showed the power of teamwork. When you're facing a long-term outage, having backup heating options and a tight-knit support system can get you through.

These real-life stories offer invaluable lessons and insights we need in order to face different scenarios with confidence. Whether it's having a well-stocked bug-out bag, reinforcing your home, or building a strong community network, these strategies can really help.

We've spent a lot of time focusing on preparing for tough situations, but now it's time to shift gears. After the crisis, how do you start picking up the pieces and rebuilding your life?

REBUILDING AFTER A DISASTER

What if a massive storm has swept through your town, leaving behind fallen trees, missing roofs, and a landscape that feels unrecognizable? It's completely normal to feel overwhelmed in the midst of all that chaos. But taking it one step at a time can make the rebuilding process feel more manageable and help you stay safe. In this chapter, we'll guide you through the steps to take after a disaster, from assessing the damage to creating a solid plan for long-term recovery.

23.1 PRACTICAL STEPS TO REBUILD AFTER A DISASTER

Assessing Damage and Safety

First things first, before you even think about stepping onto your property to assess the damage, make sure you've got the all-clear from local authorities. Your safety should always come first. When you're getting set to take a look, don't forget to put on protective gear: gloves, sturdy boots, and a helmet.

First, check the structural integrity of your home. Look for visible signs of damage, for example, cracks in the walls, sagging ceilings, or a shifted foundation. Inspect the roof for damage, too. Keep an eye out for potential hazards like gas leaks and electrical issues; if you smell gas or notice frayed wires, don't hesitate to call in the professionals. Remember, debris can hide dangers, so move cautiously as you navigate through the area.

Here's a sample checklist for a post-disaster damage assessment of your home:

1. Check for gas leaks first. If you smell gas or hear any hissing, shut off the main valve right away and call in professionals. Gas leaks are a serious risk and need to be dealt with immediately.
2. Take a look at your electrical system. Before you think about turning the power back on, check for any frayed wires or sparks. If you notice damage, keep the power off and get an electrician in to handle it.
3. Clear any major debris. Focus on clearing walkways first, but be careful. Debris can hide sharp objects or unstable areas. It's important to have safe paths before diving into the rest of the assessment.
4. Inspect the roof. Look out for missing shingles, leaks, or any sagging spots that could get worse if left alone.
5. Check the structure of your home. Walk around and see if any walls are leaning, floors are buckling, or if doors and windows aren't closing right. These could be signs of bigger issues with the structure.
6. Examine plumbing. Look for burst pipes or leaks, and make sure sinks and toilets are working as they should. Water damage can sneak up on you if left unchecked.
7. Look at the foundation. Cracks, shifts, or any settling might mean serious trouble with the stability of your home later, so keep an eye out for those.
8. Watch for water damage. Stains, swelling, or mold on the walls and ceilings are signs that water has made its way somewhere, and that's something you'll want to fix quickly.

- Check for pests. After a disaster, rodents or insects might have found their way inside and can cause damage and pose health risks, so it's worth giving the place a good check.

Prioritizing Rebuilding Tasks. Image courtesy of Digital Proshots

After you've assessed the damage, it's time to figure out what needs to be done first. Focus on restoring the essentials: getting the roof, walls, and foundation back in shape should be your priority.

- Having a secure shelter is vital for both safety and comfort.
- Next, focus on restoring your water supply and waste disposal systems.Having clean water and proper waste management is essential for staying healthy during the recovery phase.
- Finally, make sure to repair and reconnect your energy systems for light and heat. Whether you rely on solar panels or a generator, having power is essential for lighting, heating, and staying connected.

Sourcing Materials and Resources

Sourcing materials for rebuilding can be tricky, but there are effective strategies to make it easier. Start by salvaging usable materials from the debris: wood, bricks, and metal can often be repurposed. Checking out local community groups and online marketplaces may lead to affordable options. Connecting with neighbors might reveal extra supplies or tools they're willing to share, which can save you time and money.

Community resource sharing is incredibly helpful; pooling tools and materials can go a long way. Consider reaching out to local rebuilding organizations for valuable support and resources, and don't hesitate to seek aid from relief organizations or government assistance programs, as they are there to offer financial help and additional workforce.

Long-Term Resilience Planning

As you rebuild, it's essential to incorporate long-term resilience planning to withstand future disasters. Materials like reinforced concrete and metal roofing are durable and can better withstand extreme weather. Design your home for disaster resistance by elevating structures in flood-prone areas and installing flood barriers. Creating redundancy in your systems is also crucial. There are multiple water sources, like wells and rainwater collection, as well as backup energy systems, such as solar panels and generators. Consider the case of a homestead design that includes elevated living spaces, flood barriers, and a combination of solar and wind energy systems for maximum resilience.s.

Rebuilding after a natural disaster is a daunting task, but with careful planning and a methodical approach, it is possible to restore your home and ensure it's better set up to withstand future risks.

23.2 ADAPTING TO NEW NORMALS

Surviving a disaster isn't just about physical strength; it also takes a toll on your mental health. It's important to be aware of the psychological effects that can linger over time. You might notice signs like irritability, insomnia, or persistent worry cropping up. Catching these feelings early on can help you handle them better. Plus, dealing with loss and change means working through the stages of grief: denial, anger, bargaining, depression, and acceptance. Each stage requires patience and self-compassion. Maintaining hope and motivation can be challenging, but it's essential for resilience. Consider personal stories of resilience, like a friend who lost everything in a wildfire but rebuilt his life by focusing on small, daily achievements and seeking community support.

Routine and structure provide stability in uncertain times. Establish daily schedules balancing work, rest, and leisure time. Setting both short-term and long-term goals gives a sense of purpose. Maintaining social connections through regular communication with family and community fosters emotional support. For example, a homestead family's daily routine could involve mornings spent tending the garden, afternoons dedicated to homeschooling kids, and evenings for family bonding around a campfire. This structure not only keeps everyone busy but also nurtures a sense of normalcy and connection.

Image courtesy of Arthouse Studio

Mindfulness and relaxation techniques are vital for managing stress. Simple meditation and deep breathing exercises, (which we will talk about in more detailin the next chapter) can significantly reduce anxiety. Journaling is a great way to record your thoughts and experiences, helping you process emotions. Incorporating physical activities like walking, yoga, or bodyweight exercises into your routine can also boost both your physical health and mental well-being. If you prefer, you can also try a simple guided meditation to help you center yourself. There are plenty of online options for guided meditations.

Educational strategies post-disaster are crucial for community engagement. Organize quick, practical sessions on essential survival skills such as water purification, first aid, and safe food handling. Community skill pooling leverages all the unique talents within the group, creating a diverse and robust learning environment. Utilize portable resources like printed manuals, solar-powered e-readers, and hand-crank radios to facilitate ongoing learning if you don't have access to infrastructure. Establish a community library to promote continuous education. Structured daily lessons on topics of interest will maintain a sense of normalcy and foster long-term resilience. Focus on youth education through hands-on activities that teach a mix of both educational topics and the immediate survival skills they need, while keeping them engaged and psychologically supported.

Entertainment solutions are vital for maintaining morale. Low-tech options like travel games, board games, card games, and books offer essential distractions. DIY entertainment, such as crafting, storytelling, and making simple musical instruments, can pass the time creatively. Turn survival skills like fire-starting or knot-tying into fun group challenges.

Mental stimulation through puzzles, riddles, and trivia keeps minds engaged, preventing boredom. Organize talent shows, skits, or group discussions for community bonding.

Mental exercises and brain games are essential for maintaining mental acuity. Use memory exercises like memorizing maps, lists, or sequences to sharpen recall skills. Engage in critical thinking puzzles, logic problems, and strategy games to enhance decision-making abilities. Use downtime to learn or practice new languages, which is useful in diverse survival situations. Regularly tackle scenarios requiring creative problem-solving, such as resource management or survival planning. Encourage team based brain games like chess and checkers to foster collaboration and strategic thinking.

If you are already engaged in heavy physical work, trying to rebuild, you won't have the energy to work out. But if not, you can absolutely maintain your physical health, and without any equipment. Keep up a routine of push-ups, squats, lunges, and planks to build strength and conditioning. Try doing three sets of each with a minute of rest in between. For cardio, running, hiking, or skipping ropes can really get your heart pumping. Consider alternating running and walking on different days each week. To stay flexible, add some stretching, yoga, or dynamic movements into your routine. A good approach is to focus on different muscle groups with daily stretching sessions.

Survival fitness drills like carrying heavy objects, climbing, or obstacle courses simulate real-world challenges. Encourage regular movement throughout the day, such as walking or manual labor, to stay active. For kids, movement breaks could include activities like short runs or mini obstacle courses. Organize group workouts or fitness challenges to keep morale high and encourage teamwork.

23.3 COMMUNITY RESILIENCE: STRENGTHENING LOCAL NETWORKS

Building a resilient community starts with cooperation and shared responsibility. In extreme situations, this may mean navigating moral dilemmas, such as deciding how to share supplies or protect others. Healthy communication within community groups fosters respect and understanding, while constructive conflict resolution helps maintain harmony. For example, a small town in the Midwest successfully navigated a severe flood by pooling assets and holding weekly meetings to discuss progress and challenges.

Resource sharing and mutual aid are fundamental to community resilience. Pooling items like tools, seeds, food, and water ensures everyone has access to essential supplies. Skill sharing is equally important. Skill-building workshops on topics like first aid, gardening, and renewable energy can be conducted to improve the community's capabilities constantly. Simulated scenarios are another effective training tool. Teaching and learning essential skills within the community will empower everyone.

Establishing community projects like shared gardens and energy systems can further solidify this network. A community resource-sharing agreement template might include sections for listing available equipment, contact information for resource owners, and guidelines for borrowing and returning supplies.

Organizing community drills and training sessions enhances disaster planning. Regular emergency drills for scenarios like fires, earthquakes, and evacuations help everyone know what to do when disaster strikes. This is particularly relevant after an earthquake when aftershocks could still potentially cause harm. Practicing responses to potential threats helps identify weaknesses and improve strategies. An outline for a community emergency drill could include roles for coordinators, specific tasks for participants, and evaluation criteria to assess the drill's effectiveness.

A sample community resilience plan template could include sections for risk assessment, a resource inventory, emergency contact lists, and detailed action steps across different types of disasters. Taking this organized approach ensures roles are clear and resources are available in times of crisis.

23.4 RESOURCE VAULT

Equipping yourself with the right reading materials and tools can make a massive difference when it comes to rebuilding after a disaster. For comprehensive guidance on structural repairs and safety, Rebuilding After the Storm: A Step-by-Step Guide to Post-Disaster Restoration by John Smith is invaluable. This book offers detailed instructions on assessing damage, sourcing materials, and prioritizing tasks, making it a must-have for anyone facing the daunting task of reconstruction.

Online sources are equally crucial. The Federal Emergency Management Agency (FEMA) website offers extensive information on disaster recovery, including downloadable checklists and guides. Their section on building offers practical tips and best practices for fortifying your home against future disasters.

Another excellent online source is the National Institute of Standards and Technology (NIST) Community Resilience Planning Guide, which helps communities prioritize and allocate assets effectively to manage risk and enhance resilience.

When it comes to tools, having a reliable set of basic equipment is essential. The DEWALT 20V MAX Combo Kit includes a range of power tools like drills and saws, which are indispensable for various rebuilding tasks. For those needing to salvage and repurpose materials, a portable generator like the Honda EU2200i can provide reliable power to run your tools when grid electricity isn't available.

For mental and emotional resilience, The Resilient Mind: Strategies for Coping with Disaster and Adapting to New Normals by Dr. Emily Brown offers deep insights into managing stress and maintaining mental well-being. This book includes personal stories of resilience, practical exercises for developing routines and structures, and techniques for mindfulness and relaxation. It's designed to help you navigate the psychological challenges of long-term survival situations.

For community engagement and educational strategies post-disaster, consider Community Resilience: A Practical Guide to BuildingStronger Networks by Mike Johnson, which is useful for anyone looking to enhance their community's ability to withstand and recover.It provides detailed advice on organizing community drills, establishing resource-sharing networks, and creating comprehensive community resilience plans.

By utilizing these suggestions, you'll be well-equipped to tackle the challenges of rebuilding and ensure your home and community are set for whatever comes next.

Rebuilding after a disaster brings its own kind of chaos, and it's easy to overlook how deeply it affects our mental health. The stress and uncertainty can be overwhelming, creating a whirlwind of emotions as you try to navigate some major adjustments. Let's explore how to focus on mental resilience during this challenging time.

MENTAL AND EMOTIONAL PREPAREDNESS

> The greatest weapon against stress is our ability to choose one thought over another.
>
> WILLIAM JAMES

Remember the chaos of a city in lockdown, deserted streets, empty shelves, and the constant hum of anxiety in the air. It's a lot to take in, and that mental strain can feel like a heavy emotional burden.

Throughout this book, we've already covered the importance of physical prepping, but let's not forget about mental and emotional health. That's what keeps us calm and focused when everything feels chaotic. So, let's get into how you can build and maintain that resilience, not just to survive but to truly thrive in any crisis.

24.1 BUILDING EMOTIONAL RESILIENCE: EXERCISES AND STRATEGIES

When tough times hit, emotional resilience is what helps you keep your cool and think straight. It's that ability to adapt and bounce back, learning from the bumps along the way. Not only does this strength help you manage your own stress, but it also allows you to be a source of support for your loved ones, helping everyone feel a bit safer when the world feels chaotic.

I can't help but think of my friend Bob when I hear about wildfires. He went through a tough time when his home was wiped out, and for weeks, he and his family had to stay in a shelter. But he was able to keep himself mentally strong so he could offer his family the support they needed, and they were able to start picking up the pieces.

Developing a Growth Mindset

> When everything seems to be going against you, remember that the airplane takes off against the wind, not with it.
>
> HENRY FORD

One of the most effective ways to build emotional resilience is by developing a growth mindset and by seeing challenges as opportunities to grow rather than insurmountable obstacles. It's about learning from failures and setbacks, not being defeated by them.

One technique is to reframe negative thoughts. Instead of thinking, "I can't handle this," change it to, "This is tough, but I can find a way through." This nudges a shift in perspective that can impact your outlook and foster mental resilience. By focusing on what you can learn and how you can grow, you turn every challenge from a massive stumbling block into a small stepping stone.

Resilience-Building Exercises

Engaging in specific exercises can really boost your emotional resilience. If this isn't something you usually do, don't worry. Even if you're a skeptic, give it a try and trust the process. Pick one exercise first and aim to practice it regularly, ideally every day, even for just a few minutes, paying attention to any shifts in your mood or behavior over time.

Journaling is a particularly powerful tool for self-reflection and emotional processing. Spend a few minutes each day, either in the morning or before bed, writing freely about your thoughts, feelings, and experiences. This practice helps you sort through your emotions and gain clarity.

- If you don't know where to start, here are some journaling prompts to help you:
- Write about a memory when you overcame a significant challenge. What strengths did you discover in yourself?
- List three things you appreciated today. How do they impact your life positively? Reflect on a recent setback. What can you learn from it, and how can it help you grow?

Gratitude practices are another effective strategy. For example, take time each day to note three things you're grateful for. This is so quick and simple, but can really foster a positive outlook and help you focus on the good, even during tough times.

Daily affirmations, like "I am capable" and "I am resilient," can really reinforce your mental strength and keep you motivated during these challenging periods.

24.2 SPECIFIC STRESS MANAGEMENT TECHNIQUES

Recognizing stress is the first step to effectively managing it. During emergencies, stress can manifest in various ways. Physically, you might experience a rapid heartbeat, sweating, or difficulty breathing. These symptoms are your body's natural response to danger, but they can escalate to stress, irritability, anxiety, and feelings of hopelessness if not managed properly. Recognizing these signs early on can help you take steps to curb your stress before it escalates.

Breathing and relaxation techniques are really helpful for managing stress.

Deep breathing exercises can really ease anxiety by inhaling through your nose deeply and exhaling through your mouth slowly.

Try Box breathing:
- Inhale for four seconds.
- Hold for four seconds.
- Exhale for four seconds.
- Hold again for four seconds.

The 4-7-8 method is also well worth trying; inhale for four seconds, hold your breath for seven, and exhale over eight seconds.

These techniques can help manage your body's stress reactions:

- Mindfulness meditation, where you focus on the here and now without any judgment, can be especially effective in soothing your nervous system.
- Guided imagery, which involves visualizing a peaceful scene, can be particularly soothing.
- Use progressive muscle relaxation by tensing and then relaxing muscle groups, gradually working through the body from head to toe, to help release physical tension.

Incorporating mindfulness and meditation into your everyday routine can provide ongoing stress management.

Basic mindfulness exercises, such as a body scan or mindful eating, keep you grounded. When body scanning, mentally "scan" your body from head to toe, paying attention to sensations in each area, one body part at a time, noticing any tension, discomfort, or relaxation. The purpose is to become more aware of physical sensations without trying to change them, helping you tune into your body and reduce stress. For mindful eating, pay full attention to the experience of eating, noticing the taste, texture, and smell of food, as well as your body's hunger and fullness cues. Instead of eating mindlessly when distracted, you slow down and savor each bite. This practice can help with portion control, improving digestion, and fostering a healthier relationship with food. It's about being present and truly appreciating the eating experience.

-Meditation practices like focused attention or loving-kindness meditation can help maintain clarity and emotional balance. The goal of focused attention is to bring your mind back to that point of focus whenever it wanders. Over time, it helps improve concentration and mindfulness. With loving-kindness meditation, you begin by generating positive feelings for yourself, and then extend these to loved ones, acquaintances, and even people you may have conflicts with. The idea is to expand your sense of compassion and develop kindness and emotional well-being. It's a great way to nurture empathy and connection to others.

Using a meditation app can be a practical way to develop a consistent practice. Apps like Headspace or Calm offer guided sessions that fit into any schedule and make it easier to integrate mindfulness into everyday life.
Developing a stress management plan tailored to your needs is crucial.

1. Start by identifying your personal stress triggers and coping mechanisms.
2. Incorporate regular relaxation practices into your daily schedule.
3. Physical activity is also essential in managing stress and improving mental health.
4. Combine your preferred strategies into your existing weekly schedules you include dedicated time for stress-relief activities. If you are able to, set aside time each morning for journaling or a brief meditation session, and a short workout. In the evenings, practice deep breathing exercises or a body scan with progressive muscle relaxation for a few minutes before going to bed. This structured approach ensures you prioritize your mental well-being, equipping you to handle emergencies with a clear, focused mind.

24.3 RESOURCE VAULT

When it comes to building mental and emotional resilience, having the right reading material at your fingertips can help you feel relaxed and in control. They can provide valuable insights and practical advice that you can bring into your daily life. Let's look at some highly recommended sources to help you enhance your mental readiness.

Another great read is Man's Search for Meaning by Viktor E. Frankl, which explores how finding purpose can help you endure even the most challenging situations.

Websites are invaluable for ongoing learning and support. The Centers for Disease Control and Prevention (CDC) offers a comprehensive guide on coping with disasters and traumatic events (https://emergency.cdc.gov/coping/index.asp), providing detailed information on maintaining emotional health during emergencies, helping you think clearly and protect yourself and your family. It also offers specific support for adults, families, children, and community leaders.

Another valuable site is Off Grid Survival (https://offgridsurvival.com), which focuses on building prepper communities and the benefits of shared assets, collective defense, and emotional support. This site also offers practical advice on forming your own community and establishing effective communication protocols.

Meditation apps are fantastic tools for incorporating mindfulness into your daily routine. Headspace is highly recommended for its variety of guided sessions suitable for beginners and advanced practitioners alike. It offers a rich library of meditations, including those focused on anxiety, self-esteem, and grief.

Calm is another great option, especially if you enjoy ambient soundscapes and sleep stories. It features guided and unguided meditations, extensive breathing exercises, and a robust section for kids.

For those on a budget, the Healthy Minds Program offers a free app with well-structured courses focusing on awareness, connection, insight, and purpose.

Smiling Mind is also free and particularly effective for kids and young adults, making it a great choice for families.

Investing in tools that support your mental and emotional health is a wise decision. A good journal, like the Leuchtturm1917 or Moleskine, can be a daily companion for self-reflection and emotional processing. Writing down your thoughts and feelings not only helps you process emotions but also provides a record of your journey, allowing you to see your growth over time.

For guided imagery and relaxation, consider a high-quality pair of noise-canceling headphones, such as the Bose QuietComfort 35 II. These can help create a peaceful environment, whether you're meditating, engaging in guided imagery, or simply trying to unwind.

Bringing these strategies into your daily routine can really boost your mental and emotional resilience. The knowledge and strategies you'll pick up from these books, websites, and apps will help you manage stress and stay balanced. As we close this chapter, keep in mind that mental and emotional resilience is an ongoing journey:the skills you build now will support you in any situation in the future.

We're getting ever closer to the finish line, where we'll go over some frequently asked questions and wrap up our Prepper's Survival Bible. But before that, let's focus on staying primed for anything, ensuring you and your family have what they need to survive.

BE EVER READY

Here's another scenario to picture yourself in: you're in the kitchen, glancing out the window as dark clouds start to roll in. You are filled with concern for your family, pets, and neighbors. When disaster hits, being able to adapt becomes crucial for getting through it. Preparing for multiple scenarios isn't just about having the right gear; it's about developing the flexibility and mindset to handle any challenge that comes your way.

In this chapter, we will revisit some of the planning you did in Chapter 1, but now that you have more knowledge and skills, you can fine-tune your plans with even more depth and spot any gaps so that these can become the next steps in your readiness plan.

25.1 PREPARING FOR MULTIPLE SCENARIOS: ADAPTABILITY AND FLEXIBILITY

Thinking ahead and preparing for curveball scenarios has huge benefits. It helps you stay primed for whatever eventualities crop up. This way, you become even more adaptable and resourceful.

Consider multiple different angles as you plan. For instance, what if a wildfire heads your way from the north instead of the south? How does that change your evacuation plan? Or what if some family members aren't home when trouble strikes, and you can't reach them?

Being flexible is essential; a rigid plan can be a stumbling block and pose risks if things don't go as expected. So, treat your plans as living documents: available for updates and improvements as you learn and adapt.

Again, this all makes sense on paper, but thinking it through and putting everything down in a plan is something that 99% of people won't do. You, with a prepper mindset, will be well set for anything because you will have thought through every eventuality and done the work.

Adapting to different possible scenarios also involves navigating the ethical dilemmas that arise during extreme situations. Picture this scenario: a severe pandemic has disrupted supply chains, and resources are scarce. You've got a pantry full of supplies, and then you notice your neighbors are having a tough time. It's a classic dilemma: how much do you share? Do you give a little to help your neighbors or share more to truly make an impact, even if it means stretching your own supplies? It's a tough decision, and finding the right balance is key to supporting both your family and your neighbors. Considering these types of ethical dilemmas ahead of time can help you make clearer choices when the situations arise.

Consider a practical example: taking both natural disasters and human threats into account. Your emergency plan should include provisions for evacuations due to hurricanes or wildfires and contingencies for civil unrest. This means having go-bags ready, understanding evacuation routes, and securing your home against intruders.

It's also essential to have communication strategies in place, ensuring you can stay in touch with family members and access reliable information. By considering various scenarios, you can create a comprehensive plan that covers a wide range of possibilities, enhancing your overall readiness.

Incorporating regular exercises and drills into your emergency planning is crucial for ensuring your plans hold up in real-life situations, such as fire evacuations or shelter-in-place orders.

Running family drills for different scenarios helps strengthen your plan and reveal any gaps that might have slipped under the radar. Think of it as a practice run to see just how prepared you really are. After each drill, take a moment to review how it went with everyone. What felt easy? What were the hiccups? By discussing it together, you can pick up some useful insights that will make your plans even better, so you're covered no matter what comes your way.

It's also wise to learn from real-life events: watching how near-misses or recent natural disasters turn out in your area can also provide insights into new strategies. For instance, if a recent flood revealed a new flood route that disrupts road access in a new location, use that information to adjust your strategy. Being adaptable and open to new lessons is key so you're positioned for the unexpected.

Additionally, engaging in community response planning and activities can provide valuable insights and foster a sense of common solidarity. Sharing knowledge and supplies with your neighbors can strengthen your collective resilience, making everyone better equipped for any emergencies.

Scenario-Based Planning Exercise

Here is an exercise to help you expand your planning:

1. Identify Potential Scenarios: List the top five emergencies most likely to affect your area (for example, hurricanes, pandemics, civil unrest).
2. Assess Your Crises Readiness: For each scenario, evaluate your current preparedness level and identify any gaps or areas for improvement.
3. Develop a Response Plan: Outline specific actions you and your family will take in each scenario. Include steps for evacuation, communication, and resource management.
4. Identify any Possible Eventualities: Identify the possible alterations you will need to factor in for each scenario.

Practice and Review:

- Conduct drills for at least two scenarios.
- After each drill, review what went well and what needs adjustment.
- Gather feedback from others where possible.
- Update your plans accordingly.

By using an adaptable and flexible mindset, you can be ready for any situation. Remember, it's not just about having the right tools and supplies;it's about being mentally and emotionally primed to face whatever comes your way.

25.2 REGULARLY UPDATING YOUR EMERGENCY PLANS

Picture this: you've put together the perfect emergency plan, but a year later, your family has grown, and new buildings have popped up around the neighborhood. Suddenly, your plan feels out of date. That's why it's so important to review and update your emergency plans regularly. Situations can change quickly: what used to be a reliable escape route might now be blocked by new construction. Keeping your plan up to date ensures it stays effective and adapts to your changing needs.

You'll need to conduct a thorough review of all components of your plan. Check for outdated information and areas that need improvement. So, if there have been changes in your family size or health conditions, update your plan, as new babies, elderly relatives, or even pets will require additional considerations. Revising evacuation routes based on new road construction or changes in local infrastructure can also make a significant difference. Regular reviews help identify these gaps and address them before they become critical issues.

Ensuring you have the proper documentation and communicating updates to others are equally important. It is good practice to keep both written and digital copies of your updated plan. Ensure all family members know about the changes and understand their roles. Holding a family meeting to review and practice the revised plan will reinforce these changes and make sure everyone is on the same page. This practice not only enhances your readiness but also fosters a sense of security and confidence among your family.

25.3 Continuous Skill Building: Workshops and Training

Think about the last time you had to rely on a skill you hadn't used in a while, maybe starting a fire in the rain or finding your way without a GPS. There's a real sense of confidence that comes from keeping your skills sharp and knowing you can handle whatever is thrown your way. Regularly refreshing and enhancing your abilities ensures you stay effective when it counts. For example, staying current with the latest first aid and survival practices allows you to adapt to new techniques and tools, making you more resilient in emergencies.

Participating in relevant workshops and training programs is a great way to go about this. Start by researching local and online sources for skill-building opportunities. Evaluate the credibility and quality of training providers by checking reviews and asking for recommendations. Joining a wilderness survival course, for example, can teach you advanced outdoor skills that you might not otherwise learn on your own.

Participating in community drills and exercises is another excellent way to enhance your survival readiness. Practicing with others not only ensures a coordinated response but also provides opportunities to network and share valuable knowledge with fellow preppers. For example, taking part in a neighborhood earthquake drill not only refines your own skills but also builds a feeling of community and mutual support, where everyone benefits from shared expertise.

Self-directed learning and practice are equally important. Set up regular practice sessions to hone your skills. Spend an afternoon in your backyard practicing fire-starting techniques or shelter-building. Utilize this book and other books or online sites we have mentioned for independent study, focusing on survival, homesteading, and all the aspects of crisis planning we have discussed. Trusted websites, online courses, and video tutorials offer a wealth of information, and the benefit is that you get to see the techniques in action. Don't forget to engage with online communities like Reddit's r/preppers to learn from others' experiences. This approach allows you to further improve and adapt on an ongoing basis.

I can't stress enough how looking ahead and staying informed about future trends is essential. Technological advances in practices, tools, equipment, clothes, energy production, and water filtration can significantly improve your survival planning. Keep an eye on expert opinions and predictions to stay ahead of the curve. For instance, emerging technologies might make survival gear more efficient or introduce new methods for sustainable living. By staying vigilant, you can ensure you and your family are fully up to date.

25.4 RESOURCE VAULT

As we wrap up this chapter and come close to the end of our journey, remember that staying equipped for challenges is an ongoing process. When it comes to being ever-ready, having access to key information makes all the difference. All the books, websites, and tools we have mentioned in the Resource Vaults offer a wealth of knowledge that can enhance your resilience and adaptability. But it is important to regularly update your knowledge, skills, and plans to ensure that you and your family can face any situation with confidence. Let's explore a couple more valuable options to keep you learning:

If you're looking to expand your skills, consider enrolling in workshops or training programs. The Boulder Outdoor Survival School offers comprehensive courses in fire-making, advanced navigation, and other key survival skills. These practical hands-on experiences can be invaluable in real-world situations.

Online options, such as the Essential Wilderness Survival Skills Online Course from Alderleaf Wilderness College, can also provide an excellent way to keep learning new skills from the comfort of your home.

When newcomers start out researching the topic of disaster planning, it's normal to feel overwhelmed or unsure. That's why the next section tackles some of the most frequently asked questions. It will also serve as a handy, reassuring round-up of some of the key lessons we have covered.

FREQUENTLY ASKED QUESTIONS

You're in the heart of a bustling prepper expo, surrounded by booths packed with survival gear, food storage kits, and people from all walks of life. The air hums with questions, conversations, and the occasional demo on how to start a fire or build an off grid solar setup. Even the most seasoned survivalists are swapping tips and seeking advice, and it's a reminder that no one has all the answers. This chapter tackles some of the most common questions, offering practical, no-nonsense advice.

What Are the Basics I Should Focus on First When Starting to Prep?

If you're just getting started, stick to the fundamentals: water, food, shelter, and security. Start by putting together a basic emergency plan: decide how you'll evacuate, where you'll meet up, and how to stay in contact if you get separated. Keep stocked up with a three-day supply of water and long-lasting food for everyone in your home. Invest in a solid water filtration system, a basic first aid kit, and essential tools like a flashlight and a multi-tool. For shelter, consider portable options like tents or emergency bivvies, and make sure you've got what you need to stay warm and dry. These essentials aim to give you a solid foundation on which to build. Security involves both physical security,such as reinforced doors and windows, and personal safety, like self-defense training and having non-lethal defense tools like pepper spray.

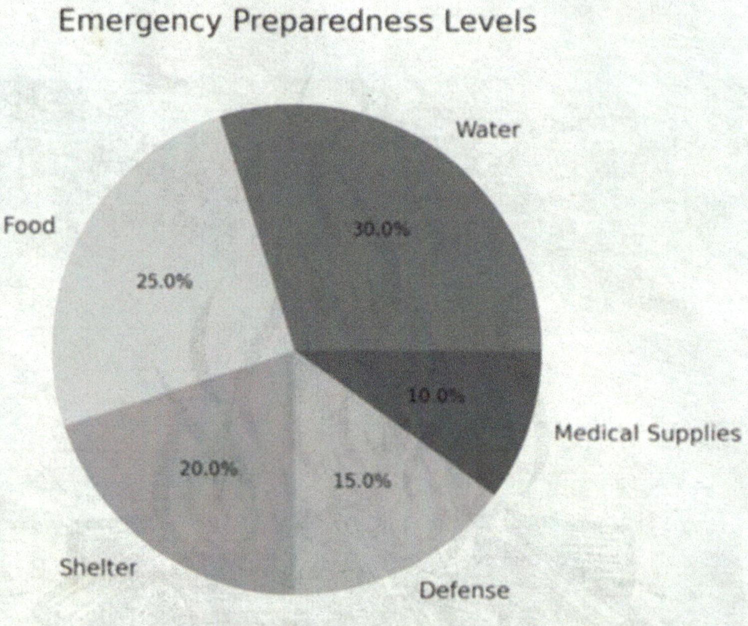

GRAPH 6: EMERGENCY PREPAREDNESS LEVELS
A BREAKDOWN OF THE FOCUS AREAS OF EMERGENCY PREPAREDNESS: WATER, FOOD, SHELTER, DEFENSE, AND MEDICAL SUPPLIES. THIS SHOWS HOW TO ALLOCATE RESOURCES IN SURVIVALIST PLANNING.
DATA COMPILED FROM STANDARD PREPAREDNESS FRAMEWORKS AND SURVIVALIST BEST PRACTICES

WHAT ARE SOME COMMON MISTAKES NEW PREPPERS MAKE?

New survivalists often make several common mistakes that can delay or hinder their response planning efforts. One mistake is focusing too much on gear and neglecting skills. It's essential to balance acquiring supplies with learning practical skills like first aid, fire starting, and food preservation. Another mistake is failing to rotate food and water supplies, leading to expired or spoiled items. Overlooking the importance of a comprehensive emergency plan is another common error. Ensure you have a detailed plan that covers communication, evacuation, and sheltering. Finally, avoid going at it alone; building a community network can significantly enhance your long-term resilience.

How Can I Store Food for Long-Term Preparedness?

Storing food for the long term requires careful planning and proper storage techniques. Aim for foods with long shelf lives, such as rice, beans, pasta, canned goods, and freeze-dried foods. Store these items in a cool, dry place, away from sunlight and pests. Use airtight containers or Mylar bags with oxygen absorbers to extend shelf life. Rotate your food stock regularly; use the oldest items first and replenish your supplies. It's also wise to include a variety of foods to ensure balanced nutrition and to keep your meal options interesting. Don't forget to stockpile spicesand condiments to enhance the flavor of your meals.

What Are the Most Important Skills to Learn for Off-Grid Living?

Living off-grid requires a diverse set of skills to ensure self-sufficiency and resilience. First and foremost, learn how to source your own food through gardening and foraging. Upgrade your skills and knowledge on soil health, planting cycles, and pest control to ensure a steady food supply. Next, acquire basic carpentry and repair skills to maintain your dwelling and build necessary structures. Water collection and purification are critical, so familiarize yourself with rainwater harvesting and filtration methods. Additionally, upskill in basic medical skills, such as first aid and wound care, to handle emergencies when professional help isn't available.

How Do I Protect My Home and Family During Civil Unrest?

Protecting your home and family involves both physical security measures and strategic planning. Start by reinforcing entry points with sturdy locks, security doors, and window bars. Install a reliable, comprehensive security system with cameras and motion sensors to monitor your property. Create a safe room within your home where you can retreat if needed. Develop a family emergency plan that addresses communication methods, meeting points, and roles for each family member. Stay informed about local events and avoid areas where unrest is likely to occur. Additionally, consider joining a local neighborhood watch program to enhance community security.

What Should Be Included in a Bug-OutBag?

A well-prepared bug-out bag should contain all the essentials you'll need to survive for at least 72 hours. Start with water and water purification methods, such as a portable filter and purification tablets

Stock up on high-energy, long-lasting foods like protein bars, nuts, and dried fruits: perfect for when you need quick fuel. Don't forget to put together a solid first aid kit, including over-the-counter meds, bandages, antiseptics, and any personal medical items you might rely on. Add a multi-tool, flashlight, extra batteries, fire starting tools, a compact sleeping bag or emergency blanket, and a change of clothes. Don't forget important documents, cash, a map, and a communication device like a two-way radio. Customize your bag based on your own specific needs and the climate of your region.

How Do I Ensure My Pet's Safety During an Emergency?

Create a pet emergency kit including water, food, medications, a leash, a collar with ID tags,and any necessary comfort items like a favorite toy or blanket.

Make sure to keep a copy of your pet's medical records and a recent photo handy, just in case you get separated. Plan ahead by finding pet-friendly shelters and hotels,so you're set if an evacuation becomes necessary. Practice evacuating with your pet to ensure they are comfortable with their carrier or crate. If you must shelter in place,designate a safe area in your home for your pet to stay secure and calm.

What Are the Best Practices for Water Filtration and Storage?

Water is crucial for survival, so having a reliable filtration and storage system is essential. Store water in clean, food-grade containers with tight seals to keep it safe. Aim for a minimum of one gallon per person per day to cover all your needs, such as drinking, cooking, and basic hygiene. Use multiple methods to purify water, such as boiling, water purification tablets, or using a portable or DIY water filter. Consider investing in a high-quality gravity water filter for home use, or making a your own, which can handle larger quantities. Regularly rotate your stored water supply and keep your filtration equipment well-maintained. Familiarize yourself with local water sources and have a plan for accessing and purifying water if your primary supply runs out.

How Do I Stay Informed During a Crisis?

Staying informed during a crisis is crucial for making timely and effective decisions. Invest in a reliable hand-crank or solar-powered radio that can receive AM, FM, and weather band signals. Subscribe to local emergency alert systems to receive real-time updates on your phone. Use social media platforms and local news websites to stay up to date on developments. Establish a communication plan in advance with family and neighbors, including regular check-ins and predetermined meeting points. Having multiple sources of information ensures you stay informed even if one method fails.

What Are Some Effective Home Defense Strategies?

Keeping yourself and your family safe begins with securing your property. Install high-quality locks on all doors and windows, and consider adding deadbolts for extra security. Use motion-activated lights all around your property to deter intruders. Install a robust security system with cameras and alarms to alert you to any suspicious activity. Create a safe room within your home where you and your family can retreat if necessary. Get training and train family members in basic self-defense techniques and consider having non-lethal defense tools like pepper spray or a stun gun readily available.

How Can I Build a Community Network for Mutual Aid?

Connect with similar-minded individuals and nurture a sense of cooperation and trust. Talk with your neighbors about how mutual aid can benefit everyone during emergencies.

If they're open to it, set up regular meetings to make plans together, run neighborhood drills, and share useful information, resources, and skills. Develop a community emergency plan that includes communication strategies, resource sharing, and roles for each member. Consider forming a neighborhood watch group to enhance security and keep everyone informed about potential threats. Building strong community ties ensures a collective response to emergencies, making everyone more resilient.

What Are the Key Components of an Effective Emergency Plan?

An effective emergency plan should include several key components: communication, evacuation, and sheltering. Start by establishing a communication plan with primary and secondary methods for staying in touch with family members. Designate a central meeting point in case of separation and ensure everyone knows how to get there. Develop an evacuation plan that specifies multiple routes and destinations, considering flexibility for different types of emergencies. Identify safe areas within your home and community where you can shelter in place if needed. Regularly review and practice your plan with all family members to ensure everyone is familiar with their roles and responsibilities.

How Do I Maintain My Prepping Supplies?

Maintain your emergency stockpile and equipment through regular checks and updates to ensure everything is in good condition. Store food in cool, dry places and rotate your stock regularly, remembering to use the oldest items first. Check the expiration dates on all your supplies and replace items as needed. Inspect your water storage containers for leaks or contamination, and refresh items now and then. Test your equipment, for example, generators, water filters, and communication devices, to make sure they are working properly. Keep an inventory of your supplies and update it regularly to track what you have and what you need to replenish.

How Do I Prepare for Economic Instability?

Preparing for economic instability involves diversifying your assets and building financial resilience. Create a budget that includes a dedicated disaster fund and a fund for unexpected emergencies. Stockpile essential items like food, water, and medical supplies to reduce reliance on fluctuating markets. Invest in precious metals like gold and silver as a hedge against currency devaluation. Build a barter inventory of high-demand items that can be traded during economic downturns. Consider learning new skills that can provide additional income or trade value.Having a financial safety net and a diverse set of resources ensures you can weather economic challenges.

How Can I Ensure My Family's Mental Health During a Crisis?

Ensuring your family's mental health during a crisis involves providing emotional support and maintaining a sense of normalcy. Encourage open communication and allow family members to express their feelings and concerns. Create daily routines that include meals, exercise, and relaxation. Incorporate mindfulness or stress-reduction techniques like deep breathing or meditation. Set up a calming space for family time and relaxation. If needed, seek professional help and remind everyone it's okay to ask for support. Taking care of mental health helps your family stay robust and equipped for whatever comes their way.

CONCLUSION

We've covered a lot of ground in this book, and I hope you now see how crucial it is to be prepared for any crisis scenario and that you feel confident to take the next steps in your own readiness. Whether it's a natural disaster, civil unrest, or economic instability, being well-equipped can make all the difference between chaos and order for you and your family.

Throughout the chapters, we've laid a solid foundation covering many of the important topics you will need to know, with additional recommendations for books, online sources, and gear. We began by exploring the mindset needed for effective planning, emphasizing the importance of self-reliance, resilience, and proactive planning. We moved on to emergency medical planning, where we discussed assembling a comprehensive first-aid kit and handling common injuries and illnesses.

Food storage and preservation was another key area, covering everything from building a prepper pantry to canning, dehydrating, and curing meat. Gardening provided a sustainable source of fresh produce, while water filtration and collection methods ensured you always have access to clean water.

We also examined renewable energy systems, including solar panels and wind turbines, to keep your home powered off the grid. We talked about battery storage systems for their role in maintaining a reliable energy supply. Non-renewable generators offer a backup for when renewable options fall short.

Home defense and self-defense strategies were addressed to keep you and your loved ones safe, followed by a look at constructing safe rooms and shelters. Urban survival tactics provided solutions for city dwellers, while bug-out bags ensured you're ready to evacuate at a moment's notice.

We also covered wilderness survival skills, from foraging for wild edibles to hunting and trapping. We covered temporary shelters and fire-starting techniques to keep you warm and protected in the wild, as well as communication plans to ensure you stay connected during emergencies. Then, we wrapped up by discussing the importance of including your family in crisis planning.

Each chapter reinforced essential lessons. These actionable steps and gear recommendations are designed to make you more self-sufficient and resilient. We also touched on the emotional and psychological aspects of resilience, ensuring you and your family can stay calm and collected in crises.

Now, it's time to take action. Don't wait for disaster to strike before you start planning. Take action today by implementing the strategies and tips you've learned. Start small if you need to, but make consistent progress. Download your free bonuses and get going. Set up that emergency kit, stockpile those essential supplies, and create that family communication plan. Practice the skills we've discussed, from fire-starting to first aid, so they become second nature.

On a personal note, I genuinely appreciate you taking the time to read this book. My passion for helping others comes from my own experience and knowledge over the years. I hope the guidance and tools here have been helpful. Remember, disaster planning is a continuous journey: it takes dedication, flexibility, and an open mindset that's always ready to learn. Stay committed, keep building your skills, and never stop refining your plans.

Lastly, thank you. Thank you for choosing this book and for taking those important steps to protect yourself and your family. Your commitment to readiness will bring peace of mind and help you stay strong, no matter what comes your way. You have the power to create a safer, more secure future for your loved ones. Stay prepared, stay safe, and remember: you've got this.

WE VALUE YOUR FEEDBACK!

Thank you for taking the time to read The Prepper's Survival Bible. Your thoughts and experiences matter to us, and we'd love to hear how this book has helped you in your preparedness journey.

Leaving a review on Amazon not only helps other readers discover this guide but also allows us to continue providing valuable content tailored to your needs.

Please take a moment to leave a review on Amazon. If you're a Goodreads user, we'd love to see your thoughts there as well. Your support makes such a difference!
Let's keep growing our community and empowering others to join us.

https://authorhelp.uk?azr=1917501013

Your thoughts matter to us
Thank you!

UNLOCK YOUR BONUSES HERE!

Congratulations on completing this next step in your survival training. As a special thank you for investing in your preparedness skills, we've gathered together all our exclusive bonuses here just for you as handy downloadables. They are all designed to help you continue to stay ready and resilient in any situation and are editable to suit your needs.

Don't miss out: unlock your bonuses now!

ACCESS ALL 15 BONUSES HERE

SCAN ME!

ACCESS EACH BONUS SEPERATELY

Bonus #1 Emergency Preparedness

SCAN ME!

Bonus #2 Printable Threat Matrix

SCAN ME!

Bonus #3 Family Communication Plan

SCAN ME!

Bonus #4 First Aid Checklist

SCAN ME!

Bonus #5 Bug out Bags - Essential and Extended

SCAN ME!

Bonus #6 Car Maintenance and Readiness Checklist

SCAN ME!

Bonus #7 Food Preservation Guide

SCAN ME!

Bonus #8 Water Filtration and Purification Cheat Sheet

SCAN ME!

Bonus #9 Gardening for Self-Sufficiency

SCAN ME!

Bonus #10 Survival Knots Cheat Sheet

SCAN ME!

Bonus #11 Barter List

SCAN ME!

Bonus #12 Spring Foraging

SCAN ME!

Bonus #13 Urban Gardening Tips

SCAN ME!

Bonus #14 Hydroponics Cheat Sheet

SCAN ME!

Bonus #15 Aquaponics Cheat Sheet

SCAN ME!

REFERENCES AND RESOURCES

BOOKS

- Aguirre, Fernando 'Ferfal'. Surviving the Economic Collapse. Independently Published, 2009.
- Alton, Joseph,and Amy Alton. The Survival Medicine Handbook. 3rd ed. Doom and Bloom, 2016.
- Baker, David L. The Complete Guide to Home Security: How to Protect Your Home and Family from Burglars, Thieves, and Other Intruders. Lyons Press, 2013.
- Beard, Daniel Carter. Shelters, Shacks, and Shanties. Dover Publications, 2008. Boxwell, Michael. The Solar Electricity Handbook. Greenstream Publishing, 2020.
- Brown, Emily.The Resilient Mind: Strategies for Coping with Disaster and Adapting to New Normals. New York: Resilience Press, 2023.
- Canterbury, Dave.Bushcraft 101: A Field Guide to the Art of Wilderness Survival. Adams Media, 2014.
- Chiras, Dan. The Homeowner's Guide to Renewable Energy: Achieving Energy Independence Through Solar,Wind, Biomass and Hydropower. 2nd ed. New Society Publishers, 2017.
- Cobb, Jim. Build the Perfect Bug-Out Bag: Your 72-Hour Disaster Survival Kit. Betterway Home, 2012.
- Cobb, Jim. The Prepper's Communication Handbook: Life-Saving Strategies for Staying in Contact During and After a Disaster. Ulysses Press, 2016.
- Cobb, Jim. The Prepper's Financial Guide: Strategies to Invest, Stockpile and Protect Your Savings for When the Banking System Collapses. Ulysses Press, 2015.
- Cobb, Jim. The Prepper's Guide to Bug-Out Bags: Your Ultimate Bug-Out Bag Packing Guide. Ulysses Press, 2016.
- Collins, Gary. Building Off the Grid: A Practical Guide to Living in Harmony with Nature.
- Second Nature Publishing, 2015.
- Del Porto, David, and Carol Steinfeld. The Composting Toilet System Book: A Practical Guide to Choosing, Planning and Maintaining Composting Toilet Systems. 3rd ed. The Center for Ecological Pollution Prevention, 2000.
- Frankl, Viktor E. Man's Search for Meaning. Beacon Press, 2006.
- Hemenway, Toby. Gaia's Garden: A Guide to Home-Scale Permaculture. 2nd ed. Chelsea Green Publishing, 2009.
- Hunt, Scott.The Prepper's Guide to Off the Grid Power: The Beginner's Guide to Solar, Wind, and Generator Systems for Remote and Emergency Use. Ulysses Press, 2016.
- Jenkins, Joseph. The Humanure Handbook: A Guide to Composting Human Manure. 4th ed. Jenkins Publishing, 2019.
- Johnson, Rich. The Ultimate Survival Manual. Weldon Owen, 2012. Kallas, John. Foraging Laws and Regulations. Self-Published, 2012.
- Kallas, John. Edible Wild Plants: Wild Foods from Dirt to Plate. Gibbs Smith, 2010.
- Kemp, William H. The Renewable Energy Handbook: A Guide to Rural Energy Independence, Off-Grid and Sustainable Living. 2nd ed. Aztext Press, 2009.
- Kingry, Judi ,and Lauren Devine. Ball Complete Book of Home Preserving: 400 Delicious and Creative Recipes for Today. Robert Rose, 2012.

- Kjellstrom, Bjorn. Be Expert with Map and Compass. Hoboken, NJ: Wiley, 2010. Korwin, Alan. Gun Owner's Guide to State Laws. Bloomfield Press, 2016.
- Luther, Daisy. The Prepper's Water Survival Guide: Harvest, Treat, and Store Your Most
- Vital Resource. Ulysses Press, 2015.
- Martens, Michael.Emergency Radio Communications: How to Build, Set Up and Operate Your Radio System for Emergencies. Kindle Direct Publishing, 2019.
- Merriam, Dwight. The Complete Guide to Zoning: How to Navigate the Complex and Expensive Maze of Zoning, Planning, Environmental and Land-Use Law. McGraw-Hill Education, 2004.
- Pennington, Tess.The Prepper's Blueprint: The Step-By-Step Guide To Help You Through Any Disaster. Survival Life, 2014.
- Pop, Lacho. Solar Power for Beginners: How to Design and Install the Best Solar Power
- System for Your Home. Self-Published, 2016.
- Rawles, James Wesley. Patriots: Surviving the Coming Collapse.Ulysses Press, 2009.
- Reivich, Karen, and Andrew Shatte.The Resilience Factor: 7 Keys to Finding Your Inner Strength and Overcoming Life's Hurdles. Broadway Books, 2003.
- Skousen, Joel. The Secure Home: Survival and Rebuilding Plans for Shelter, Security, and Sustenance. 3rd ed. Self-Published, 2007.
- Smith, Edward C. The Vegetable Gardener's Bible. 2nd ed. Storey Publishing, 2009.
- Smith, John.Rebuilding After the Storm: A Step-by-Step Guide to Post-Disaster Restoration. Self-Published, 2015.
- Southern, Adrian, and Whelm King. The AquaponicFarmer: A Complete Guide to Building and Operating a Commercial Aquaponic System. New Society Publishers, 2017.
- Spigarelli Aston,Patricia. Crisis Preparedness Handbook:A Comprehensive Guide to Home Storage and Physical Survival. Cross-Current Publishing, 2016.
- Stewart, Creek. Build the Perfect Bug Out Bag: Your 72-Hour Disaster Survival Kit.
- Betterway Home, 2012.
- Stone, Curtis. The UrbanFarmer: Growing Food for Profit on Leased and Borrowed Land.
- New Society Publishers, 2015.
- Thayer, Samuel.Forager's Harvest: A Guide to Identifying, Harvesting, and Preparing Edible Wild Plants. Forager's Harvest Press, 2006.
- Wise, Barbara. Container Gardening for All Seasons: Enjoy Year-Round Color with 101 Designs. Cool Springs Press, 2012.
- Wise, Scott. Krav Maga: An Essential Guide to the Renowned Method of Self-Defense. St.
- Martin's Griffin, 2014.

WEBSITES

- American Radio Relay League (ARRL). Website for amateur radio enthusiasts and emergency communications. www.arrl.org.
- Animated Knots by Grog. AccessedSeptember 27, 2024. https://www.animatedknots.com.
- Backpacking Chef. Rehydrating Foods. https://www.backpackingchef.com/rehydratingfoods.html.
- Battery University. A complete guide to battery technology and maintenance. www.batteryuniversity.com.
- California Residential Mitigation Program. Resource for strengthening homes to resist earthquakes. www.earthquakebracebolt.com.

- Centers for Disease Control and Prevention. Accessed September 27, 2024. https://www.cdc.gov.
- Clean Energy Reviews. Solar energy reviews and comparisons. www.cleanenergyreviews.info.
- Crest Real Estate. Real estate and development services. www.crestrealestate.com.
- EnergySage.Comparison of solar energy systems. www.energysage.com.
- Epic Gardening. Gardening tips and guides for all types ofgrowers. www.epicgardening.com.
- Family Handyman "Diesel vs. Propane Generators." Accessed September 27, 2024. https://www.familyhandyman.com.
- Federal Emergency Management Agency (FEMA). Disaster preparedness and recovery resources. www.fema.gov.
- FEMA Safe Rooms for Tornadoes and Hurricanes. Guide to building safe rooms. www.fema.gov/safe-rooms.
- Fortified Estate. High-security solutions for homes. www.fortifiedestate.com.
- Gardening Off The Grid. Sustainable gardening for off-grid living. www.gardeningoffthegrid.com.
- Goldmoney. App for buying and storing precious metals. www.goldmoney.com.
- Investopedia. Precious Metals Investing Guide for Beginners. www.investopedia.com.
- LegalMatch. Find lawyers for various legal issues. www.legalmatch.com.
- Missouri Department of Conservation. Conservation and outdoor resources for Missouri residents. www.mdc.mo.gov.
- MIRA Safety. Professional-grade gas masks and other safety equipment. www.mirasafety.com.
- National Center for Home Food Preservation. Expert advice on home canning and food preservation. nchfp.uga.edu.
- National Institute of Standards and Technology (NIST) Community Resilience Planning Guide. Guide for improving community resilience. www.nist.gov.
- Off-Grid Survival. Survivalist and prepping information for off-grid living. www.offgridsurvival.com.
- Old Farmer's Almanac Companion Planting Chart. Chart for effective companion planting. www.almanac.com.
- Outdoor Life. Articleson hunting, fishing, and outdoor survival. www.outdoorlife.com.
- Pantry Check. App for managing and organizing your food pantry. www.pantrycheck.com.
- PlantSnap.App for identifying plants using your phone camera. www.plantsnap.com.
- Ready.gov. Government preparedness resources. www.ready.gov.
- Red Cross. Emergency services and disaster response.www.redcross.org.
- Rob Bob's Aquaponics and Backyard Farm. YouTube channel for aquaponics and sustainable farming. www.youtube.com/RobBob.
- Scouting Magazine - Finding Water in the Wilderness. Article on finding water in survival situations. www.scoutingmagazine.org.
- Security.org Article: Best DIY Home Security Systems of 2024. https://www.security.org/home-security-systems/best/
- Security.org Article: Complete Guide to Home Security Systems. https://www.security.org/home-security-systems/
- SurvivalSherpa. Blog focused on traditional survival skills and self-reliance. www.survivalsherpa.com.
- The Bug Out Bag Guide. Guides and tips for building bug-out bags. www.thebugoutbagguide.com.
- The Prepared. Comprehensive survival guides for modern emergencies. www.theprepared.com.
- Unbound Solar. Accessed September 27, 2024. https://unboundsolar.com.
- Wilderness Awareness School. Courses and resources for wilderness survival. www.wildernessawareness.org.
- Windustry.org. Wind power information for consumers. www.windustry.org.

- YNAB (You Need A Budget). App for personal budgeting and finance management. www.ynab.com.
- ZoningCheck. Tool for checking zoning regulations in different areas. www.zoningcheck.com.

WEBSITES

- American Radio Relay League (ARRL). Website for amateur radio enthusiasts and emergency communications. www.arrl.org.
- Animated Knots by Grog. AccessedSeptember 27, 2024. https://www.animatedknots.com.
- Backpacking Chef. Rehydrating Foods. https://www.backpackingchef.com/rehydratingfoods.html.
- Battery University. A complete guide to battery technology and maintenance. www.batteryuniversity.com.
- California Residential Mitigation Program. Resource for strengthening homes to resist earthquakes. www.earthquakebracebolt.com.

ADDITIONAL SOURCES

- 3 Critical Maintenance Tips to Keep Your Crossbow in Top Shape. TenPoint Crossbows. https://www.tenpointcrossbows.com/blog/3-critical-maintenance-tips-to-keep-your-crossbow-in-top-shooting-form/
- 4x4 Gears and Gadgets for Your Car. https://www.4x4australia.com.au/gear/electronics-gadgets/4x4-gear-and-gadgets
- 5 Key Steps to Making a Crisis-Proof Food Supply. Allnewshub. https://allnewshub.co.uk/5-key-steps-to-making-a-crisis-proof-food-supply/
- 7 Best Emergency Water Filters for Survival. WaterFilterGuru. https://waterfilterguru.com/best-emergency-water-filter/
- 10 Best Non Lethal Self Defense Weapons. The Home Security Superstore Blog. https://www.thehomesecuritysuperstore.com/blogs/the-home-security-superstore-blog/10-best-non-lethal-self-defense-weapons
- 10 Disaster Survival Tips for Pets. MapQuest Travel. https://www.mapquest.com/travel/10-disaster-survival-tips-for-pets.htm
- 10 Methods to Preserve Wild Edibles. Slow Living Kitchen. https://slowlivingkitchen.com/preserving-wild-edibles/
- 10 Techniques for Catching Fish in a Survival Situation. Outdoor Life. https://www.outdoorlife.com/blogs/survivalist/10-techniques-catching-fish-survival-situation/
- 12 of the Best Non-Perishable Foods. Healthline. https://www.healthline.com/nutrition/non-perishable-food
- 20 Edible Weeds You Can Forage From Your Back Yard. Almanac. https://www.almanac.com/edible-weeds-garden-eating-weeds
- 20 Prepper FAQs and Q&A. TruePrepper. https://trueprepper.com/prepper-faqs-qa/
- 20 Wild Edible Plants. Farmers' Almanac. https://www.farmersalmanac.com/20-common-wild-edible-plants
- A Complete Guide on How to Install a Solar Panel at Home. Skills Training Group. https://www.skillstg.co.uk/blog/how-to-install-a-solar-panel/
- A Forager's Checklist for Safe, Sustainable & Ethical Foraging. Gather+ Grow. https://gatherandgrow.com/blog/a-foragers-checklist-for-safe-sustainable-amp-ethical-foraging
- A Guide to the 15 Best Survival Traps of All Time. Outdoor Life. https://www.outdoorlife.com/how-build-trap-15-best-survival-traps/

ADDITIONAL SOURCES

- 3 Critical Maintenance Tips to Keep Your Crossbow in Top Shape. TenPoint Crossbows. https://www.tenpointcrossbows.com/blog/3-critical-maintenance-tips-to-keep-your-crossbow-in-top-shooting-form/
- 4x4 Gears and Gadgets for Your Car. https://www.4x4australia.com.au/gear/electronics-gadgets/4x4-gear-and-gadgets
- 5 Key Steps to Making a Crisis-Proof Food Supply. Allnewshub. https://allnewshub.co.uk/5-key-steps-to-making-a-crisis-proof-food-supply/
- 7 Best Emergency Water Filters for Survival. WaterFilterGuru. https://waterfilterguru.com/best-emergency-water-filter/
- 10 Best Non Lethal Self Defense Weapons. The Home Security Superstore Blog. https://www.thehomesecuritysuperstore.com/blogs/the-home-security-superstore-blog/10-best-non-lethal-self-defense-weapons
- 10 Disaster Survival Tips for Pets. MapQuest Travel. https://www.mapquest.com/travel/10-disaster-survival-tips-for-pets.htm
- 10 Methods to Preserve Wild Edibles. Slow Living Kitchen. https://slowlivingkitchen.com/preserving-wild-edibles/
- 10 Techniques for Catching Fish in a Survival Situation. Outdoor Life. https://www.outdoorlife.com/blogs/survivalist/10-techniques-catching-fish-survival-situation/
- 12 of the Best Non-Perishable Foods. Healthline. https://www.healthline.com/nutrition/non-perishable-food
- 20 Edible Weeds You Can Forage From Your Back Yard. Almanac. https://www.almanac.com/edible-weeds-garden-eating-weeds
- 20 Prepper FAQs and Q&A. TruePrepper. https://trueprepper.com/prepper-faqs-qa/
- 20 Wild Edible Plants. Farmers' Almanac. https://www.farmersalmanac.com/20-common-wild-edible-plants
- A Complete Guide on How to Install a Solar Panel at Home. Skills Training Group. https://www.skillstg.co.uk/blog/how-to-install-a-solar-panel/
- A Forager's Checklist for Safe, Sustainable & Ethical Foraging. Gather+ Grow. https://gatherandgrow.com/blog/a-foragers-checklist-for-safe-sustainable-amp-ethical-foraging
- A Guide to the 15 Best Survival Traps of All Time. Outdoor Life. https://www.outdoorlife.com/how-build-trap-15-best-survival-traps/
- Are you prepared to save a life? Take our CPR Quiz. Canadian Red Cross. https://www.redcross.ca/quiz/are-you-prepared-to-save-a-life-take-our-cpr-quiz
- Assessing Structural Damage After A Natural Disaster. Oklahoma State University Extension. https://extension.okstate.edu/programs/emergency-and-disaster-preparedness/flooding/assessing-structural-damage-after-a-natural-disaster.html
- Atlas SurvivalShelters. https://atlassurvivalshelters.com/
- Batteries vs. Generators: How to Choose the Best Backup Power Solution. CNET. https://www.cnet.com/home/solar/batteries-or-generators-choosing-the-best-home-backup-power-solution/
- Best Budgeting Apps Of September 2025. Forbes Advisor. https://www.forbes.com/advisor/banking/best-budgeting-apps/
- Best DIY Home Security Systems of 2025. Security.org. https://www.security.org/home-security-systems/best/diy/

- Best Fire Starters of 2025, Tested and Reviewed. Outdoor Life. https://www.outdoorlife.com/gear/best-fire-starter/
- Best Home Solar Battery Systems. Clean Energy Reviews. https://www.cleanenergyreviews.info/blog/best-solar-battery-systems
- Best Non-Lethal Self-Defense Weapons [2023]. Athlon Outdoors. https://athlonoutdoors.com/article/non-lethal-self-defense-options/
- The 5 Best Survival Schools: Learn Skills for Any Scenario. GearJunkie. https://gearjunkie.com/outdoor/best-survival-schools
- Best Survival Jackets:Synthetic and Natural. TruePrepper. https://trueprepper.com/survival-jacket/
- Best Survival Knives of 2024,Tested and Reviewed. Outdoor Life. https://www.outdoorlife.com/gear/best-survival-knife/
- Best Whole House Generators Of September 2024. Forbes Home. https://www.forbes.com/home-improvement/electrical/best-whole-house-generators/
- Best Bug Out Bag to Survive 2025: The Definitive Guide. MIRA Safety. https://www.mirasafety.com/blogs/news/2024-bug-out-bag-gear-guide?srsltid=AfmBOorlIxiWeFE_ISdXOM-oPR5dcQJJLGtw_-QVflQo14kdo3MkRhTG
- Bug Out Bag List of Essentials. Tips from Expert . Mountain house.com https://mountainhouse.com/blogs/emergency-prep-survival/bug-out-bag
- Build a Pet Disaster Preparedness Kit. Centers for Disease Control and Prevention. https://www.cdc.gov/healthy-pets/emergency-preparedness/preparedness-kit.html
- Building a disaster 'grag and go' bag for your pet BCAA. https://www.bcaa.com/blog/membership/how-to-make-a-disaster-grab-and-go-bag-for-your-pet
- Car Emergency Kits 101 [Guide]. Allstate. https://www.allstate.com/resources/car-insurance/car-emergency-kit-list
- Community Resilience Planning Guide. National Institute of Standards and Technology.https://www.nist.gov/community-resilience/planning-guide
- Companion Planting Chart and Guide for Vegetable Gardens. The Old Farmer's Almanac. https://www.almanac.com/companion-planting-guide-vegetables
- Coping with a Disaster or Traumatic Event. Centers for Disease Control and Prevention. https://emergency.cdc.go v/coping/index.asp
- Defensible Space. Cal Fire. https://www.fire.ca.gov/dspace
- Diesel vs. Propane Generators. Family Handyman. https://www.familyhandyman.com/article/diesel-vs-propane-generators/
- Disaster And Its Impact On Mental Health: A Narrative Review. Journal of Family Medicine and Primary Care. https://www.ncbi.nlm.nih.gov/pmc/articles/PMC6857396/
- Disaster Preparedness and Response. National Institute of Environmental Health Sciences. https://tools.niehs.nih.gov/wetp/index.cfm?id=556
- Disaster Preparedness in and Recovery in Older Adults. National Institute of Aging https://www.nia.nih.gov/health/safety/disaster-preparedness-and-recovery-older-adults
- DIY Solar Food Dehydrator Plans. Mother Earth News. https://www.motherearthnews.com/diy/solar-food-dehydrator-plans-zm0z14jjzmar/
- Do It Yourself Retrofit Tips for California Homes. California Residential Mitigation Program. https://www.earthquakeauthority.com/prepare-your-house-earthquake-risk/brace-and-bolt-grants/cea-policyholder-brace-bolt-grants/homeowner-resources/do-it-yourself-tips
- Emergency Shelter Guide: Choosing and Using the Right Shelter for Outdoor Survival. Defiance Gear Co. https://defiancegear.com/blogs/news/emergency-shelter-guide-choosing-and-using-the-right-shelter-for-outdoor-survival

- Ethical Dilemmas in Disaster Medicine. Western Journal of Emergency Medicine. https://www.ncbi.nlm.nih.gov/pmc/articles/PMC3518976/
- Finding Water in the Wilderness. BC Outdoors. https://www.bcoutdoorsmagazine.com/finding-water-in-the-wilderness/
- First Aid Training. American Red Cross. https://www.redcross.org/take-a-class/first-aid/first-aid-training
- Free Crisis Communication Plan Templates. Smart sheet. https://www.smartsheet.com/content/crisis-communication-templates
- Gardening Australia (2022). How to make your own solar food dehydrator. YouTube. https://www.youtube.com/watch?v=Z0f0jew8Whw
- Gardening Off The Grid. An Off Grid Life. https://www.anoffgridlife.com/gardening-off the-grid/
- Generator Safety Tips. American Red Cross. https://www.redcross.org/get-help/how-to-prepare-for-emergencies/types-of-emergencies/power-outage/safe-generator-use.html
- Get help during an emergency with your Android phone. Google Support. https://support.google.com/android/answer/9319337?hl=en
- Getting Licensed.ARRL: The National Association for Amateur Radio. http://www.arrl.org/getting-licensed
- Grow a Survival Garden to Prepare for Natural Disasters and Crisis. Irish Survivalist and Prepper Group. https://irish-survival-group.com/survival-garden
- Guide on Firearms Licensing Law. SCEG UK. (Downloadable) https://share.google/KbFgGY5D20uWHFOjN
- History of the Prepper Movement:From Paranoid Radicals to Mainstream Survival. History Cooperative. https://historycooperative.org/history-of-the-prepper-movement/
- Home Fire Safety Tip Sheets from the National Fire Protection Association. NFPA. https://www.nfpa.org/education-and-research/home-fire-safety/safety-tip-sheets
- Homemade DIY Grey Water System Water Recycling. Autodesk Indestructibles https://www.instructables.com/Home-made-DIY-Grey-water-system-water-recycling/
- Home Remedies for Survival. Vocal Media Education. https://vocal.media/education/home-remedies-for-survival-100-ways-to-prepare-for-the-worst
- How to Build A Safe Room In An Existing Home. Fortified Estate. https://fortifiedestate.com/retrofit-home-safe-room/
- How to Build a Fire Pit. Lowe's. https://www.lowes.com/n/how-to/how-to-build-a-custom-fire-pit
- How to Build a Snow Cave. Scout Life magazine. https://scoutlife.org/outdoors/150860/how-to-build-a-snow-cave/
- How to Build an Emergency Tarp Shelter. Outdoors.org. https://www.outdoors.org/resources/amc-outdoors/outdoor-resources/how-to-build-an-emergency-tarp-shelter/
- How to Check Zoning Laws and Ordinances in Your Area. LegalMatch. https://www.legalmatch.com/law-library/article/how-to-find-zoning-and-ordinances.html
- How to Cook Food Over a Campfire I Top Tips and Tricks. KOA.
- https://koa.com/blog/how-to-cook-food-over-a-campfire-pro-campfire-cooking-tips/
- How to Make a First Aid Kit/ Supplies & Contents. American Red Cross. https://www.redcross.org/get-help/how-to-prepare-for-emergencies/anatomy-of-a-first-aid-kit.html
- How to Make a Potting Mix. Royal Horticultural Society. https://www.rhs.org.uk/soil-composts-mulches/how-to-make-your-own-potting-mix
- How to Make an All-Natural First Aid Kit. Steroplast Healthcare. https://www.steroplast.co.uk/knowledge-base/how-to-make-an-all-natural-first-aid-kit.html

- How to Make a Survival Water Filter.Wilderness Awareness School. https://wildernessawareness.org/articles/how-to-make-a-survival-water-filter/
- How to Make Your Own Hiking Survival Kit – 10 Essential Items to Include. Live for the Outdoors. https://www.livefortheoutdoors.com/hiking/advice/how-to-make-your-own-hiking-survival-kit/
- How to perform CPR - Adults CPR Steps. American Red Cross. https://www.redcross.org/take-a-class/cpr/performing-cpr/cpr-steps
- How to Prepare For Emergencies I Be Red Cross Ready. American Red Cross. https://www.redcross.org/get-help/how-to-prepare-for-emergencies.html
- How to Start a Prepper Pantry - The Complete Guide (Free). The Revolution Blog. https://www.therevolutionblog.com/prepper-pantry/
- How to Stock Your Storm Shelter Or Safe Room. Armored Closet. https://www.armoredcloset.com/articles/how-to-stock-your-storm-shelter-or-safe-room/
- How to support your family during a crisis. UNH Extension. https://extension.unh.edu/blog/2020/03/how-support-your-family-during-crisis
- Hunter Ethics. Missouri Department of Conservation. https://mdc.mo.gov/hunting-trapping/regulations/hunter-ethics
- Hurricane Katrina: Survivors on the Storm's Impact. People. https://people.com/human-interest/hurricane-katrina-survivors-15th-anniversary/
- Krav Maga Worldwide. https://www.kravmaga.com/
- Lead-Acid vs. Lithium Batteries: Which Are Best For Solar? Unbound Solar. https://unboundsolar.com/blog/lead-acid-vs-lithium-batteries
- Lesson 3. Assessing Risks. FEMA Training. https://training.fema.gov/emiweb/is/is393a/is393.a-lesson3.pdf
- Lifesaving Training from the Leader in CPR Science and Education https://cpr.heart.org/en/
- Make A Plan. Ready.gov.https://www.ready.gov/plan
- Make a First Aid Kit/ Supplies & Contents. American Red Cross. https://www.redcross.org/get-help/how-to-prepare-for-emergencies/anatomy-of-a-first-aid-kit.html
- Make a Household Gravity Water Filter in 10 Mins. Instructables. https://www.instructables.com/Make-a-Household-Gravity-Water-Filter-in-10-Mins/
- Managing stress after a disaster/ Community support. Queensland Government. https://www.qld.gov.au/community/disasters-emergencies/disasters/mental-health/managing-stress-after
- Michigan State University. Basics of Pressure CanningFoods. Michigan State University Extension.https://www.canr.msu.edu/news/basics_of_pressure_canning_foods
- Mid-March 2019: Historical, Catastrophic Flooding Impacts Nebraska. National Weather Service. https://www.weather.gov/gid/march2019flood
- Most Efficient Solar Panels for 2024: New Record-Setting Technology. CNET. https://www.cnet.com/home/energy-and-utilities/most-efficient-solar-panels/
- Natural Garden Pest Control Methods. Marie Bostwick. https://mariebostwick.com/natural-garden-pest-control-methods/
- Navigatingthe Legalities of Building and Living Off-Grid. Reject the Grid. https://rejectthegrid.com/navigating-the-legalities-of-building-and-living-off-grid/
- Perovskite Solar Cells. U.S. Department of Energy. https://www.energy.gov/eere/solar/perovskite-solar-cells
- Physical Fitness and Medications for Camping. AcampX. https://acampx.com/physical-fitness-and-medications-for-camping/
- Planning for Recovery Before a Disaster Strikes https://coast.noaa.gov/digitalcoast/training/disaster-recovery-preparedness.html
- Portable Generators. Rainier Outdoor. https://rainieroutdoorpower.com/
- Precious Metals Investing Guide for Beginners. Investopedia. https://www.investopedia.com/terms/p/preciousmetal.asp

- Preparedness 101: The Beginner's Guide to Food Storage. Food Storage Moms. https://www.foodstoragemoms.com/preparedness-101-the-beginners-guide-tofood-storage/
- Prepper Communities: Building a Survival Network in Troubled Times. Off Grid Survival. https://offgridsurvival.com/prepper-communities-building-a-survival-network-in-troubled-times/
- Prepper Community: Strategies for Self-Reliant Survival Networks. TAC Skills.https://tac-skills.com/prepper-community-strategies/
- Prepper's Guide: 20 Essential Items to Keep in Your Bug-Out Bag. Survival Sullivan. (Downloadable) https://www.survivalsullivan.com/bug-out-bag-checklist/
- Primitive Fire Making: 6 Ways To Make Fire Without Matches. The Bug Out Bag Guide. https://www.thebugoutbagguide.com/primitive-fire-making/
- Proper Field Dressing and Handling of Wild Game and Fish. Penn State Extension. https://extension.psu.edu/proper-field-dressing-and-handling-of-wild-game-and-fish
- Psychological First Aid Training. Mental Health First Aid. https://www.mentalhealthfirstaid.org/about/
- Remote Tripwire Alarm : 7 Steps (with Pictures). Instructables. https://www.instructables.com/Remote-Tripwire-Alarm/
- Restrictions and Barriers to Renewable Energy in Local Zoning. Pacific Northwest National Laboratory. (Downloadable) https://share.google/yLIELIDT45AmOTm6k
- Safe Rooms for Tornadoes and Hurricanes. FEMA. (Downloadable) https://share.google/WrNUudGiVJkk6EH2s
- Scouting Report - Foraging for Wild Edibles. Scouting Magazine. https://scoutingmagazine.org/2016/01/foraging-for-wild-edibles/
- Soil Testing.NC StateExtension. https://gardening.ces.ncsu.edu/soils/soil-testing/
- Solar Battery Installation Explained / Step-by-Step Guide. The Eco Experts. https://www.theecoexperts.co.uk/solar-panels/solar-battery-installation
- Solar Energy DIY- Let the sun power your home. Motoma. https://motoma.com/battery/effortless-solar-a-simple-guide-to-installing-solar-energy-storage-system.html
- Solids Filter For Aquaponic System/Fish Tank Canister Filter (2018). Rob Bob's Aquaponics and Backyard Farm. https://www.youtube.com/watch?v=HtMZRM3a5Fs
- Step-by-Step Guide to Canning Foods. Midwest Living. https://www.midwestliving.com/food/step-by-step-guide-to-canning/
- Survival Gear Essentials. Urban SurvivalSite. https://www.survivalgearanswers.com/15-essential-urban-survival-gear-items/
- Survival Myths: Facts vs. Fiction. The Prepared https://www.beprepared.com/blogs/articles/fact-or-fiction-natural-disaster-survival-myths
- Testing Acidity and Alkalinity of Soil. Grow Observatory. UK.
- https://share.google/XOz6WjWu1A8Fuetby
- The 4 Best Meditation Apps of 2024 / Reviews by Wirecutter. The New YorkTimes. https://www.nytimes.com/wirecutter/reviews/best-meditation-apps/
- The 5 Best Backpacking Water Filters of 2024 / Tested. Outdoor Gear Lab. https://www.outdoorgearlab.com/topics/camping-and-hiking/best-backpacking-water-filter
- The 5 Best Satellite Messengers of 2024 / Tested. Outdoor Gear Lab. https://www.outdoorgearlab.com/topics/camping-and-hiking/best-personal-locator-beacon
- The Beginner's Guide to Foraging. Backpacker. https://www.backpacker.com/skills/foraging/
- The Best Emergency Radios in 2024, Tried and Tested. CNN Underscored. https://www.cnn.com/cnn-underscored/reviews/best-emergency-radios
- The Best Survival Gear for 2024. Trailspace. https://www.trailspace.com/gear/survival/
- The Best Weather Apps for 2025. PCMag. https://uk.pcmag.com/mobile-apps/132792/the-best-weather-apps-for-2021

- The Texas Manual on Rainwater Harvesting. Texas Water Development Board. (Downloadable) https://serc.carleton.edu/details/files/20136.html
- The Top 50 Items To Barter With In Case Of Disaster. US Preppers. https://uspreppers.com/the-top-50-items-to-barter-with-in-case-of-disaster/
- The Trailblazers of CPR and First Aid: History, Benefits, and Rescues. Health Mystery Canadian PR. https://healthmystery.ca/the-trailblazers-of-cpr-and-first-aid-history-benefits-a nd-rescues/
- The Ultimate Aquaponics Beginner's Guide. Go Green Aquaponics. https://gogreenaquaponics.com/blogs/news/ultimate-aquaponics-beginners-guide
- The Ultimate Guide to Homestead Food Preservation Strategies for 2025. Homesteading Lifestyle.
- https://homesteadinglifestyle.com/homestead-food-preservation/
- Things to Pack in Your 72-Hour EmergencyKit. Ready.gov. https://www.ready.gov/kit
- Tips for Setting Up Tents Fast. KOA. https://koa.com/blog/how-to-set-up-any-tent-fast-how-to-pitch-different-types-of-tents/
- Top 5 Best Survival Kits HiConsumption. https://hiconsumption.com/gear/best-survival-kits/
- Top 5 Off-Grid Water Purification Systems. The Review Hub.
- https://www.thereviewhub.com/off-grid-water-purification-systems/
- Top 10 Energy StorageTrends in 2023. BloombergNEF.
- https://about.bnef.com/blog/top-10-energy-storage-trends-in-2023/
- Top 10 Fire Starter Kits. Family Handyman. https://www.familyhandyman.com/list/fire-starter-kits/
- Top 10 Survival Knots You Need to Know. TruePrepper. https://trueprepper.com/survival-knots/
- Top Prepper Blog to Follow. Prepper's Will. https://bloggers.feedspot.com/survival_blogs/
- TruePrepper. (2024). Prepper Risk Assessment. https://trueprepper.com/prepper-risk-assessment/
- Understanding Urban Agriculture - What is Urban Farming? Natural Resources Conservation Service. https://www.nrcs.usda.gov/getting-assistance/other-topics/urban-agriculture
- Unleash your urban survival gardening Preppers Only https://preppersonly.co.uk/unleash-your-urban-survival-gardening
- Urban Gardens: How to Plant in a Small Space. LawnStarter.
- https://www.lawnstarter.com/blog/gardening-2/urban-gardens-small-space-planting/
- US 'Prepper' Culture diversifies Amid Fear Of Disaster And Uncertainty. Reuters. https://www.reuters.com/world/us/prepping-disaster-diversifies-more-americans-lose-trust-2024-03-09/
- Vertical Gardening for Small Spaces. Mostardi Nursery. https://mostardi.com/vertical-gardening-for-small-spaces/
- What is a Zoning Variance and When Do You Need One? CrestReal Estate. https://www.crestrealestate.com/zoning-variance/
- Wild Edible Plants. Wild Food UK. https://www.wildfooduk.com/wild-plant-guide/
- Wild EdiblePlants: A Guide to Common Edible Species. Wild Edible. https://www.wildedible.com/
- Wind Turbines vs. Solar Panels for Home - Which Is Better? Treehugger. https://www.treehugger.com/wind-turbines-vs-solar-panels-for-home-5187949

QUOTES

- Addison, J. The Spectator. London, UK: The Spectator, 1712.
- Cascio, J. The American Resilience Project. Retrieved from American Resilience Project, n.d.
- Ford, H. My Life and Work. New York, NY: Doubleday, Page & Company, 1922.
- Franklin, B. Poor Richard's Almanack. Philadelphia, PA: BenjaminFranklin, n.d.
- Francis of Assisi. Various historical texts. Quoted in multiple sources including The Little Flowers of St. Francis, n.d.
- James, W. The Principles of Psychology. New York, NY: Henry Holt and Company, 1890.
- Kimball, S. W. Faith Precedes the Miracle. Salt Lake City, UT: Deseret Book, 1976.
- King, S. Night Shift. New York, NY: Doubleday, 1978.
- Lincoln, A. Various historical records.Quoted in multiple sources including Collected Works of Abraham Lincoln, 1865.
- Mabee, M. Civil Defense Book. Retrieved from Civil Defense Book, n.d.
- Prepper Proverb. Commonly cited in prepper and survivalist communities. Retrieved from Various Prepping Sites, n.d.
- Sagan, C. Cosmos. New York, NY: Random House, 1980.
- Seneca the Younger. In Letters from a Stoic, translated by Robin Campbell. London: Penguin Books, 1969.
- Shaw, G. B. Man and Superman. London, UK: Constable and Company, 1903.

Made in United States
Orlando, FL
04 November 2026

90202950R00090